MATH USER'S HANDBOOK

hot words
hot topics

Glencoe
McGraw-Hill

New York, New York
Columbus, Ohio
Chicago, Illinois
Peoria, Illinois
Woodland Hills, California

An **Online Multilingual Glossary,** which includes
11 languages in addition to English, can be found
at **www.math.glencoe.com/multilingual_glossary**

Glencoe

The McGraw·Hill Companies

Send all inquiries to:
Glencoe/McGraw-Hill
8787 Orion Place
Columbus, OH 43240-4027

ISBN 0-07-860083-9   *Quick Review Math Handbook, Course 1*

Printed in the United States of America.
  17 18 19 20 026 10 09 08

HANDBOOK AT A GLANCE

CONTENTS

Introduction   xiv

Descriptions of features show you how to use this handbook.

PART TWO

Hot Topics   64

A reference to key topics spread over nine areas of mathematics

8 MEASUREMENT   348

xi

PART THREE

Hot Solutions and Index — 404

xiii

INTRODUCTION

Why use this handbook?

You will use this mathematics handbook to refresh your memory of concepts and skills.

What are Hot Words and how do you find them?

The Hot Words section includes a glossary of terms, a collection of common or significant mathematical patterns, and lists of symbols and formulas in alphabetical order. Many entries in the glossary will refer you to chapters and topics in the Hot Topics section for more detailed information.

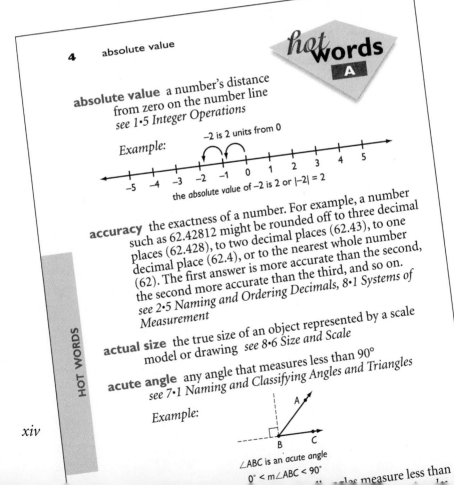

4   absolute value

*hot*words
A

absolute value a number's distance from zero on the number line
see 1•5 Integer Operations

Example:

−2 is 2 units from 0

the absolute value of −2 is 2 or |−2| = 2

accuracy the exactness of a number. For example, a number such as 62.42812 might be rounded off to three decimal places (62.428), to two decimal places (62.43), to one decimal place (62.4), or to the nearest whole number (62). The first answer is more accurate than the second, the second more accurate than the third, and so on.
see 2•5 Naming and Ordering Decimals, 8•1 Systems of Measurement

actual size the true size of an object represented by a scale model or drawing *see 8•6 Size and Scale*

acute angle any angle that measures less than 90°
see 7•1 Naming and Classifying Angles and Triangles

Example:

∠ABC is an acute angle
0° < m∠ABC < 90°

HOT WORDS

What are Hot Topics and how do you use them?

The Hot Topics section consists of nine chapters. Each chapter has several topics that give you to-the-point explanations of key mathematical concepts. Each topic includes one or more concepts. After each concept is a Check It Out section, which gives you a few problems to check your understanding of the concept. At the end of each topic, there is an exercise set.

There are problems and a vocabulary list at the beginning and end of each chapter to help you preview and review what you know.

What are Hot Solutions?

The Hot Solutions section gives you easy-to-locate answers to Check It Out and What Do You Already Know? problems.

HOT SOLUTIONS

1•4 FACTORS AND MULTIPLES

1•4 Facto

Factors

Suppose that you want rectangular pattern.
$$1 \times 15 = 15$$

$$3 \times 5 = 15$$

Two numbers multiplied by considered **factors** of 15. So t To decide whether one numbe there is a remainder of 0, the n

FINDING THE FACT

What are the factors of 20?
- Find all pairs of numbers th product.
$$1 \times 20 = 20 \qquad 2 \times 10 =$$
- List the factors in order, startii

The factors of 20 are 1, 2, 4, 5, 10,

Check It Out
Find the factors of each num
1. 6
                    2. 18

hot **words**

The Hot Words section includes a glossary of terms, a collection of common or significant mathematical patterns, and lists of symbols and formulas. Many entries in the glossary will refer to chapters and topics in the Hot Topics section.

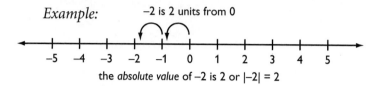

absolute value a number's distance from zero on the number line *see 1•5 Integer Operations*

Example: −2 is 2 units from 0

the *absolute value* of −2 is 2 or |−2| = 2

accuracy the exactness of a number. For example, a number such as 62.42812 might be rounded off to three decimal places (62.428), to two decimal places (62.43), to one decimal place (62.4), or to the nearest whole number (62). The first answer is more accurate than the second, the second more accurate than the third, and so on. *see 2•5 Naming and Ordering Decimals, 8•1 Systems of Measurement*

actual size the true size of an object represented by a scale model or drawing *see 8•6 Size and Scale*

acute angle any angle that measures less than 90° *see 7•1 Naming and Classifying Angles and Triangles*

Example:

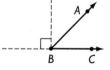

∠ABC is an *acute angle*

0° < m∠ABC < 90°

acute triangle a triangle in which all angles measure less than 90° *see 7•1 Naming and Classifying Angles and Triangles*

Example:

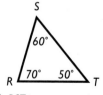

△ RST is an *acute triangle*

additive inverse a number that when added to a given number results in a sum of zero

> *Example:* $(+3) + (-3) = 0$
> (-3) is the *additive inverse* of 3

additive property the mathematical rule that states that if the same number is added to each side of an equation, the expressions remain equal

additive system a mathematical system in which the values of individual symbols are added together to determine the value of a sequence of symbols

> *Examples:* The Roman numeral system, which uses symbols such as I, V, D, and M, is a well-known additive system.

This is another example of an additive system:

$$\triangledown\triangledown\square$$

If $\square$ equals 1 and $\triangledown$ equals 7,

then $\triangledown\triangledown\square$ equals $7 + 7 + 1 = 15$

algebra a branch of mathematics in which symbols are used to represent numbers and express mathematical relationships *see Chapter 6 Algebra*

algorithm a specific step-by-step procedure for any mathematical operation *see 2·3 Addition and Subtraction of Fractions, 2·4 Multiplication and Division of Fractions, 2·6 Decimal Operations*

altitude the perpendicular distance from the base of a shape to the vertex. *Altitude* indicates the height of a shape.

> *Example:*

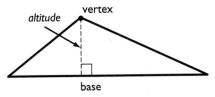

angle two rays that meet at a common endpoint
see 7·1 Naming and Classifying Angles and Triangles

Example:

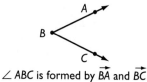

∠ ABC is formed by $\overrightarrow{BA}$ and $\overrightarrow{BC}$

angle of elevation an angle formed by an upward line of sight and the horizontal

Example:

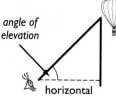

angle of elevation

horizontal

apothem a perpendicular line from the center of a regular polygon to any one of its sides

Example:

apothem

approximation an estimate of a mathematical value that is not exact but close enough to be of use

Arabic numerals (or Hindu-Arabic numerals) the number symbols we presently use {0, 1, 2, 3, 4, 5, 6, 7, 8, 9}

arc a section of a circle *see 7·8 Circles*

Example:

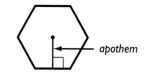

$\overset{\frown}{QR}$ is an *arc*

area the size of a surface, usually expressed in square units
see *7•5 Area, 7•6 Surface Area, 7•8 Circles, 8•3 Area,
Volume, and Capacity*

Example: 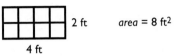   2 ft   area = 8 ft²

4 ft

arithmetic expression a mathematical relationship
expressed as a number, or two or more numbers with
operation symbols
see *6•1 Writing Expressions and Equations*

arithmetic sequence a mathematical progression in which
the difference between any two consecutive numbers in
the sequence is the same see *page 61*

Example: 2, 6, 10, 14, 18, 22, 26
the common difference of this *arithmetic
sequence* is 4

associative property a rule that states that the sum or
product of a set of numbers is the same, no matter how
the numbers are grouped
see *1•2 Properties, 6•2 Simplifying Expressions*

Examples: $(x + y) + z = x + (y + z)$
$x \times (y \times z) = (x \times y) \times z$

average the sum of a set of values divided by the number of
values see *4•4 Statistics*

Example: the *average* of 3, 4, 7, and 10 is
$(3 + 4 + 7 + 10) \div 4 = 6$

average speed the average rate at which an object moves

axis (pl. *axes*) [1] one of the reference lines by which a point on
a coordinate graph may be located; [2] the imaginary line
about which an object may be said to be symmetrical
(*axis* of symmetry); [3] the line about which an object
may revolve (*axis* of rotation) see *6•6 Graphing on the
Coordinate Plane, 7•3 Symmetry and Transformations*

bar graph a way of displaying data using horizontal or vertical bars *see 4•2 Displaying Data*

base [1] the side or face on which a three-dimensional shape stands; [2] the number of characters a number system contains *see 1•1 Place Value of Whole Numbers, 7•6 Surface Area, 7•7 Volume*

base-ten system the number system containing ten single-digit symbols {0, 1, 2, 3, 4, 5, 6, 7, 8, and 9} in which the numeral 10 represents the quantity ten *see 1•1 Place Value of Whole Numbers, 2•5 Naming and Ordering Decimals*

base-two system the number system containing two single-digit symbols {0 and 1} in which 10 represents the quantity two *see binary system*

benchmark a point of reference from which measurements can be made *see 2•7 Naming Percents*

best chance in a set of values, the event most likely to occur *see 4•6 Probability*

bimodal distribution a statistical model that has two different peaks of frequency distribution *see 4•3 Analyzing Data*

binary system the base two number system, in which combinations of the digits 1 and 0 represent different numbers, or values

binomial an algebraic expression that has two terms

   Examples: $x^2 + y$; $x + 1$; $a - 2b$

box plot a diagram, constructed from a set of numerical data, that shows a box indicating the middle 50% of the ranked statistics, as well as the maximum, minimum, and medium statistics *see 4•2 Displaying Data*

broken-line graph a type of line graph used to show change over a period of time *see 4·2 Displaying Data*

Example:

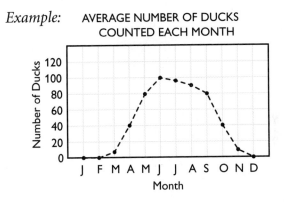

AVERAGE NUMBER OF DUCKS
COUNTED EACH MONTH

budget a spending plan based on an estimate of income and expenses *see 9·4 Spreadsheets*

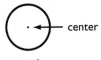

cells small rectangles in a spreadsheet that hold information. Each rectangle can store a label, number, or formula *see 9·4 Spreadsheets*

center of the circle the point from which all points on a circle are equidistant *see 7·8 Circles*

chance the probability or likelihood of an occurrence, often expressed as a fraction, decimal, percentage, or ratio *see 2·9 Fraction, Decimal, and Percent Relationships, 4·6 Probability, 6·4 Ratio and Proportion*

circle a perfectly round shape with all points equidistant from a fixed point, or center *see 7·8 Circles*

Example:

center

a *circle*

circle graph (pie chart) a way of displaying statistical data by dividing a circle into proportionally-sized "slices" *see 4·2 Displaying Data*

Example:

FAVORITE PRIMARY COLOR

circumference the distance around a circle, calculated by multiplying the diameter by the value pi *see 7·8 Circles*

classification the grouping of elements into separate classes or sets *see 5·3 Sets*

collinear a set of points that lie on the same line

Example:

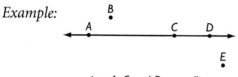

points A, C, and D are *collinear*

columns vertical lists of numbers or terms; in spreadsheets, the names of cells in a column all beginning with the same letter {A1, A2, A3, A4, . . .} *see 9·4 Spreadsheets*

combination a selection of elements from a larger set in which the order does not matter *see 4·5 Combinations and Permutations*

Example: 456, 564, and 654 are one *combination* of three digits from 4567

common denominator a whole number that is the denominator for all members of a group of fractions *see 2·3 Addition and Subtraction of Fractions*

Example: the fractions $\frac{5}{8}$ and $\frac{7}{8}$ have a *common denominator* of 8

common difference the difference between any two consecutive terms in an arithmetic sequence *see arithmetic sequence*

HOT WORDS

common factor a whole number that is a factor of each number in a set of numbers
see 1•4 Factors and Multiples

Example: 5 is a *common factor* of 10, 15, 25, and 100

common ratio the ratio of any term in a geometric sequence to the term that precedes it *see geometric sequence*

commutative property the mathematical rule that states that the order in which numbers are added or multiplied has no effect on the sum or product
see 1•2 Properties, 6•2 Simplifying Expressions

Examples: $x + y = y + x$
$x \cdot y \cdot z = y \cdot x \cdot z$

composite number a number exactly divisible by at least one whole number other than itself and 1
see 1•4 Factors and Multiples

concave polygon a polygon that has an interior angle greater than 180°

Example:

270°

a concave polygon

conditional a statement that something is true or will be true provided that something else is also true
see contrapositive, converse, 5•1 If/Then Statements

Example: if a polygon has three sides, then it is a triangle

cone a solid consisting of a circular base and one vertex

Example:

vertex

a cone

congruent figures figures that have the same size and shape. The symbol ≅ is used to indicate congruence.

Example:

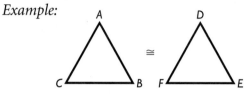

triangles *ABC* and *DEF* are *congruent*

conic section the curved shape that results when a conical surface is intersected by a plane

Example:

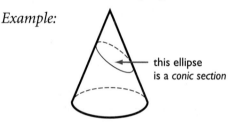

this ellipse
is a *conic section*

continuous data that relate to a complete range of values on the number line

Example: the possible sizes of apples are *continuous* data

contrapositive a logical equivalent of a given conditional statement, often expressed in negative terms
see 5•1 If/Then Statements

Example: "if *x*, then *y*" is a conditional statement;
"if not *y*, then not *x*" is the *contrapositive*
statement

convenience sampling a sample obtained by surveying people on the street, at a mall, or in another convenient way as opposed to a random sample *see 4•1 Collecting Data*

converse a conditional statement in which terms are expressed in reverse order *see 5•1 If/Then Statements*

Example: "if *x*, then *y*" is a conditional statement;
"if *y*, then *x*" is the *converse* statement

convex polygon a polygon that has no interior angle greater than 180°
see 7·2 *Naming and Classifying Polygons and Polyhedrons*

Example:

a regular hexagon is a *convex polygon*

coordinate graph the representation of points in space in relation to reference lines—usually, a horizontal *x*-axis and a vertical *y*-axis
see *coordinates*, 6·6 *Graphing on the Coordinate Plane*

coordinates an ordered pair of numbers that describes a point on a coordinate graph. The first number in the pair represents the point's distance from the origin (0, 0) along the *x*-axis, and the second represents its distance from the origin along the *y*-axis. *see ordered pairs*, 6·6 *Graphing on the Coordinate Plane*

Example:

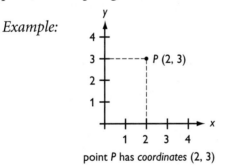

point *P* has *coordinates* (2, 3)

coplanar points or lines lying in the same plane

correlation the way in which a change in one variable corresponds to a change in another

cost an amount paid or required in payment

cost estimate an approximate amount to be paid or to be required in payment

HOT WORDS

counterexample a specific example that proves a general mathematical statement to be false
see 5•2 Counterexamples

counting numbers the set of numbers used to count objects; therefore, only those numbers that are whole and positive {1, 2, 3, 4,...}
see positive integers

cross product a method used to solve proportions and test whether ratios are equal: $\frac{a}{b} = \frac{c}{d}$ if $ad = bc$
see 6•4 Ratio and Proportion

cross section the figure formed by the intersection of a solid and a plane

Example:

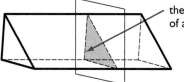

the *cross section* of a triangular prism

cube (n.) a solid figure with six square faces
see 7•2 Naming and Classifying Polygons and Polyhedrons

Example:

a *cube*

cube (v.) to multiply a number by itself and then by itself again *see 3•1 Powers and Exponents*

Example: $2^3 = 2 \times 2 \times 2 = 8$

cube root the number that must be multiplied by itself and then by itself again to produce a given number

Example: $\sqrt[3]{8} = 2$

cubic centimeter the amount contained in a cube with edges that are 1 cm in length *see 7•7 Volume*

cubic foot the amount contained in a cube with edges that are 1 foot in length *see 7•7 Volume*

cubic inch the amount contained in a cube with edges that are 1 inch in length *see 7•7 Volume*

cubic meter the amount contained in a cube with edges that are 1 meter in length *see 7•7 Volume*

customary system units of measurement used in the United States to measure length in inches, feet, yards, and miles; capacity in cups, pints, quarts, and gallons; weight in ounces, pounds, and tons; and temperature in degrees Fahrenheit *see English system, 8•1 Systems of Measurement*

cylinder a solid shape with parallel circular bases *see 7•6 Surface Area*

Example:

a *cylinder*

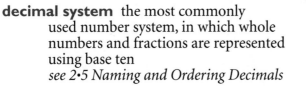

decagon a plane polygon with ten angles and ten sides

decimal system the most commonly used number system, in which whole numbers and fractions are represented using base ten *see 2•5 Naming and Ordering Decimals*

Example: decimal numbers include 1230, 1.23, 0.23, and −123

degree [1] (algebraic) the exponent of a single variable in a simple algebraic term; [2] (algebraic) the sum of the exponents of all the variables in a more complex algebraic term; [3] (algebraic) the highest degree of any term in an equation; [4] (geometric) a unit of measurement of an angle or arc, represented by the symbol °
see [1] 3•1 Powers and Exponents, [4] 7•1 Naming and Classifying Angles and Triangles, 7•8 Circles, 9•2 Scientific Calculator

Examples: [1] In the term $2x^4y^3z^2$, x has a *degree* of 4, y has a *degree* of 3, and z has a *degree* of 2.

[2] The term $2x^4y^3z^2$ as a whole has a *degree* of $4 + 3 + 2 = 9$.

[3] The equation $x^3 = 3x^2 + x$ is an equation of the third *degree*.

[4] An acute angle is an angle that measures less than 90°.

denominator the bottom number in a fraction
see 2•1 Fractions and Equivalent Fractions

Example: for $\frac{a}{b}$, b is the *denominator*

dependent events a group of happenings, each of which affects the probability of the occurrence of the others
see 4•6 Probability

diagonal a line segment that connects one vertex to another (but not one next to it) on a polygon
see 7•2 Naming and Classifying Polygons and Polyhedrons

Example:

$\overline{BD}$ is a *diagonal* of parallelogram ABCD

diameter a line segment that passes through the center of a circle and divides it in half *see 7·8 Circles*

Example:

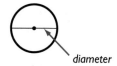

diameter

difference the result obtained when one number is subtracted from another

dimension the number of measures needed to describe a figure geometrically

Examples: A point has 0 *dimensions.*
A line or curve has 1 *dimension.*
A plane figure has 2 *dimensions.*
A solid figure has 3 *dimensions.*

direct correlation the relationship between two or more elements that increase and decrease together

Example: At an hourly pay rate, an increase in the number of hours you work means an increase in the amount you get paid, while a decrease in the number of hours you work means a decrease in the amount you get paid.

discount a deduction made from the regular price of a product or service *see 2·8 Using and Finding Percents*

discrete data that can be described by whole numbers or fractional values. The opposite of *discrete* data is continuous data.

Example: the number of oranges on a tree is *discrete* data

distance the length of the shortest line segment between two points, lines, planes, and so forth
see 8·2 Length and Distance

distance-from graph a coordinate graph that shows distance from a specified point as a function of time

HOT WORDS

distribution the frequency pattern for a set of data
see 4·3 *Analyzing Data*

distributive property of multiplication over addition
multiplication is *distributive* over addition.
For any numbers x, y, and z,
$x(y + z) = xy + xz$
see 1·2 *Properties*, 6·2 *Simplifying Expressions*

double-bar graph a graphical display that uses paired
horizontal or vertical bars to show a relationship
between data see 4·2 *Displaying Data*

Example:

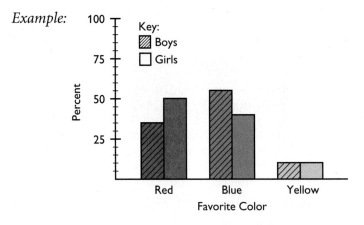

edge a line along which two planes
of a solid figure meet
see 7·2 *Naming and Classifying
Polygons and Polyhedrons*

English system units of measurement used in the United
States that measure length in inches, feet, yards, and
miles; capacity in cups, pints, quarts, and gallons;
weight in ounces, pounds, and tons; and temperature
in degrees Fahrenheit see *customary system*

equal angles angles that measure the same number of degrees
see 7·1 *Naming and Classifying Angles and Triangles*

equally likely describes outcomes or events that have the same chance of occurring *see 4•6 Probability*

equally unlikely describes outcomes or events that have the same chance of not occurring *see 4•6 Probability*

equation a mathematical sentence stating that two expressions are equal
see 6•1 Writing Expressions and Equations

Example: $3 \times (7 + 8) = 9 \times 5$

equiangular having more than one angle, each of which is the same size

equiangular triangle a triangle in which each angle is 60°
see equilateral triangle, 7•1 Naming and Classifying Angles and Triangles

equilateral a shape having more than one side, each of which is the same length

equilateral triangle a triangle in which each side is of equal length *see equiangular triangle, 7•1 Naming and Classifying Angles and Triangles*

Example:

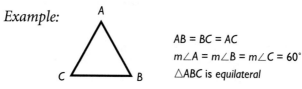

$AB = BC = AC$
$m\angle A = m\angle B = m\angle C = 60°$
$\triangle ABC$ is equilateral

equivalent equal in value

equivalent expressions expressions that always result in the same number, or have the same mathematical meaning for all replacement values of their variables
see 6•2 Simplifying Expressions

Examples: $\frac{9}{3} + 2 = 10 - 5$

$2x + 3x = 5x$

equivalent fractions fractions that represent the same quotient but have different numerators and denominators
see 2·1 Fractions and Equivalent Fractions

Example: $\frac{5}{6} = \frac{15}{18}$

equivalent ratios ratios that are equal
see 6·4 Ratio and Proportion

Example: $\frac{5}{4} = \frac{10}{8}$; 5:4 = 10:8

estimate an approximation or rough calculation

even number any whole number that is a multiple of 2
{0, 2, 4, 6, 8, 10, 12, . . .}

event any happening to which probabilities can be assigned
see 4·6 Probability

expanded notation a method of writing a number that highlights the value of each digit
see 1·1 Place Value of Whole Numbers

Example: 867 = 800 + 60 + 7

expense an amount of money paid; cost

experimental probability a ratio that shows the total number of times the favorable outcome happened to the total number of times the experiment was done
see 4·6 Probability

exponent a numeral that indicates how many times a number or expression is to be multiplied by itself

Example: in the equation $2^3 = 8$, the *exponent* is 3

expression a mathematical combination of numbers, variables, and operations; e.g., $6x + y^2$ *see 6·1 Writing Expressions and Equations, 6·2 Simplifying Expressions, 6·3 Evaluating Expressions and Formulas*

face a two-dimensional side of a
three-dimensional figure
*see 7•2 Naming and Classifying
Polygons and Polyhedrons, 7•6 Surface Area*

factor a number or expression that is multiplied by another
to yield a product *see 1•4 Factors and Multiples*

Example: 3 and 11 are *factors* of 33

factor pair two unique numbers multiplied together to yield a
product, such as 2 × 3 = 6 *see 1•4 Factors and Multiples*

factorial represented by the symbol !, the product of all the
whole numbers between 1 and a given positive whole
number *see 4•5 Combinations and Permutations*

Example: 5! = 1 × 2 × 3 × 4 × 5 = 120

fair describes a situation in which the theoretical probability
of each outcome is equal
see 4•6 Probability

Fibonacci numbers *see page 61*

flip to "turn over" a shape
see reflection, 7•3 Symmetry and Transformations

Example:

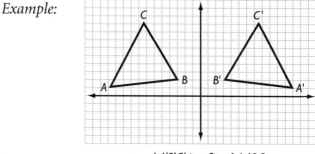

△A'B'C' is a flip of △ABC

forecast to predict a trend, based on statistical data
see 4•3 Analyzing Data

HOT WORDS

formula an equation that shows the relationship between two or more quantities; a calculation performed by spreadsheet *see pages 58–59, 6•3 Evaluating Expressions and Formulas, 9•4 Spreadsheets*

> *Example:* $A = \pi r^2$ is the *formula* for calculating the area of a circle; A2 $\times$ B2 is a spreadsheet *formula*

fraction a number representing some part of a whole; a quotient in the form $\frac{a}{b}$
see 2•1 Fractions and Equivalent Fractions

frequency graph a graph that shows similarities among the results so one can quickly tell what is typical and what is unusual *see 4•2 Displaying Data*

function assigns exactly one output value to each input value

> *Example:* You are driving at 50 mi/hr. There is a relationship between the amount of time you drive and the distance you will travel. You say that the distance is a *function* of the time.

geometric sequence a sequence in which the ratio between any two consecutive terms is the same *see common ratio and page 61*

> *Example:* 1, 4, 16, 64, 256, . . . the common ratio of this *geometric sequence* is 4

geometry the branch of mathematics concerned with the properties of figures *see Chapter 7 Geometry, 9•3 Geometry Tools*

gram a metric unit used to measure mass *see 8•3 Area, Volume, and Capacity*

hot **words** G

greatest common factor (GCF) the greatest number that
is a factor of two or more numbers
see 1·4 Factors and Multiples

Example: 30, 60, 75
the *greatest common factor* is 15

growth model a description of the way data change over time

harmonic sequence *see page 61*

height the distance from the base to the
top of a figure *see 7·7 volume*

heptagon a polygon that has seven sides

Example:

a *heptagon*

hexagon a polygon that has six sides

Example:

a *hexagon*

hexagonal prism a prism that has two hexagonal bases and
six rectangular sides

Example:

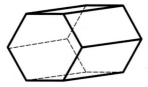

a *hexagonal prism*

HOT WORDS

hexahedron a polyhedron that has six faces

Example:

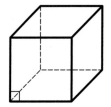

a cube is a *hexahedron*

histogram a graph in which statistical data is represented by blocks of proportionately-sized areas
see 4•2 Displaying Data

horizontal a flat, level line or plane

hypotenuse the side of a right triangle, opposite the right angle
see 7•1 Naming and Classifying Angles and Triangles

Example:

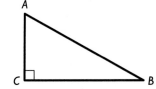

side $\overline{AB}$ is the *hypotenuse* of this right triangle

improper fraction a fraction in which the numerator is greater than the denominator *see 2•1 Fractions and Equivalent Fractions*

Examples: $\frac{21}{4}, \frac{4}{3}, \frac{2}{1}$

income the amount of money received for labor, services, or the sale of goods or property

independent event an event in which the outcome does not influence the outcome of other events *see 4•6 Probability*

inequality a statement that uses the symbols > (greater than), < (less than), ≥ (greater than or equal to), and ≤ (less than or equal to) to indicate that one quantity is larger or smaller than another *see 6•5 Inequalities*

Examples: $5 > 3$; $\frac{4}{5} < \frac{5}{4}$; $2(5 - x) > 3 + 1$

infinite, nonrepeating decimal irrational numbers, such as π and $\sqrt{2}$, that are decimals with digits that continue indefinitely but do not repeat

inscribed figure a figure that is enclosed by another figure as shown below

Examples:

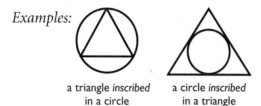

a triangle *inscribed* in a circle   a circle *inscribed* in a triangle

integers the set of all whole numbers and their additive inverses $\{\ldots -5, -4, -3, -2, -1, 0, 1, 2, 3, 4, 5 \ldots\}$

intercept [1] the cutting of a line, curve, or surface by another line, curve, or surface; [2] the point at which a line or curve cuts across a given axis

intersection the set of elements that belong to each of two overlapping sets *see 5•3 Sets*

Example:

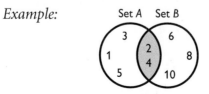

The shaded area is the *intersection* of set A (numbers 1 through 5) and set B (even numbers to 10).

inverse operations operations that undo each other

> *Examples:* Addition and subtraction are inverse
> operations: $5 + 4 = 9$ and $9 - 4 = 5$.
> Adding 4 is the inverse of subtracting by 4.
> Multiplication and division are inverse
> operations: $5 \times 4 = 20$ and $20 \div 4 = 5$.
> Multiplying by 4 is the inverse of dividing by 4.

irrational numbers the set of all numbers that cannot be
expressed as finite or repeating decimals

> *Examples:* $\sqrt{2}$ (1.414214 ...) and π (3.141592 ...) are
> *irrational numbers*

isometric drawing a two-dimensional representation of a
three-dimensional object in which parallel edges are
drawn as parallel lines

> *Example:*

isosceles trapezoid a trapezoid in which the two
nonparallel sides are of equal length

> *Example:*

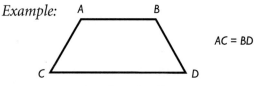

AC = BD

an *isosceles trapezoid*

isosceles triangle a triangle with at least two sides of equal
length
see 7·1 Naming and Classifying Angles and Triangles

> *Example:*

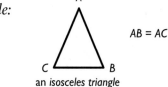

AB = AC

an *isosceles triangle*

hot **words**
L

law of large numbers when you experiment by doing something over and over, you get closer and closer to what things "should" be theoretically. For example, when you repeatedly throw a die, the proportion of 1's that you throw will get closer to $\frac{1}{6}$ (which is the theoretical proportion of 1's in a batch of throws).

leaf the unit-digit of an item of numerical data between 1 and 99

least common denominator (LCD) the least common multiple of the denominators of two or more fractions *see 2•3 Addition and Subtraction of Fractions*

Example: 12 is the *least common denominator* of $\frac{1}{3}, \frac{2}{4}$, and $\frac{3}{6}$

least common multiple (LCM) the smallest nonzero whole number that is a multiple of two or more whole numbers *see 1.4 Factors and Multiples, 2•3 Addition and Subtraction of Fractions*

Example: the *least common multiple* of 3, 9, and 12 is 36

legs of a triangle the sides adjacent to the right angle of a right triangle

Example:

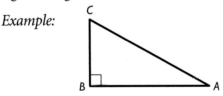

$\overline{AB}$ and $\overline{BC}$ are the *legs of triangle ABC*

length a measure of the distance of an object from end to end *see 8•2 Length and Distance*

like terms terms that include the same variables raised to the same powers. *Like terms* can be combined. *see 6•2 Simplifying Expressions*

Example: $5x^2$ and $6x^2$ are like terms; $3xy$ and $3zy$ are not like terms

likelihood the chance of a particular outcome occurring *see 4•6 Probability*

line a connected set of points extending forever in both directions
see 7•1 Naming and Classifying Angles and Triangles

line graph data displayed visually to show change over time *see 4•2 Displaying Data*

Example:

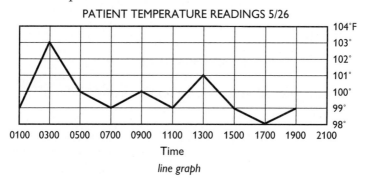

PATIENT TEMPERATURE READINGS 5/26

Time

line graph

line of symmetry a line along which a figure can be folded so that the two resulting halves match
see 7•3 Symmetry and Transformations

Example:

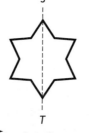

$\overleftrightarrow{ST}$ is a *line of symmetry*

line segment a section of a line running between two points
see 7•1 Naming and Classifying Angles and Triangles

Example: A •————• B
$\overline{AB}$ is a *line segment*

linear measure the measure of the distance between two points on a line

liter a basic metric unit of capacity
see 8•3 Area, Volume, and Capacity

logic the mathematical principles that use existing theorems
to prove new ones *see Chapter 5 Logic*

loss an amount of money that is lost

lowest common multiple the smallest number that is a
multiple of all the numbers in a given set; same as least
common multiple *see 1•4 Factors and Multiples*

Example: for 6, 9, and 18, 18 is the *lowest common
multiple*

Lucas numbers *see page 62*

magic square *see page 62*

mathematical argument a series of
logical steps a person might follow to
determine whether a statement is correct

maximum value the greatest value of a function or a set of
numbers

mean the quotient obtained when the sum of the numbers in
a set is divided by the number of addends
see average, 4•4 Statistics

Example: the *mean* of 3, 4, 7, and 10 is
$(3 + 4 + 7 + 10) \div 4 = 6$

measurement units standard measures, such as the meter,
the liter, and the gram, or the foot, the quart, and the
pound *see 8•1 Systems of Measurement*

median the middle number in an ordered set of numbers
see 4•4 Statistics

Example: 1, 3, 9, 16, 22, 25, 27
16 is the *median*

HOT WORDS

meter the basic metric unit of length

metric system a decimal system of weights and measurements based on the meter as its unit of length, the kilogram as its unit of mass, and the liter as its unit of capacity *see 8·1 Systems of Measurement*

midpoint the point on a line segment that divides it into two equal segments

Example:

M

A •——————•—————————• B

AM = MB

M is the *midpoint* of $\overline{AB}$

minimum value the least value of a function or a set of numbers

mixed number a number composed of a whole number and a fraction *see 2·3 Addition and Subtraction of Fractions*

Example: $5\frac{1}{4}$

mode the number or element that occurs most frequently in a set of data *see 4·4 Statistics*

Example: 1, 1, 1, 2, 2, 3, 5, 5, 6, 6, 6, 6, 8
      6 is the *mode*

monomial an algebraic expression consisting of a single term. $5x^3y$, xy, and $2y$ are three *monomials.*

multiple the product of a given number and an integer *see 1·4 Factors and Multiples*

Examples: 8 is a *multiple* of 4
      3.6 is a *multiple* of 1.2

multiplication growth number a number that when used to multiply a given number a given number of times results in a given goal number

Example: grow 10 into 40 in two steps by multiplying
(10 × 2 × 2 = 40)
2 is the *multiplication growth number*

multiplicative inverse the number for any given number
that will yield 1 when the two are multiplied, same as
reciprocal

Example: $10 \times \frac{1}{10} = 1$
$\frac{1}{10}$ is the *multiplicative inverse* of 10

natural variability the difference in
results in a small number of
experimental trials from the
theoretical probabilities

negative integers the set of all integers that are less than zero

Examples: $-1, -2, -3, -4, -5, \ldots$

negative numbers the set of all real numbers that are less
than zero

Examples: $-1, -1.36, -\sqrt{2}, -\pi$

net a two-dimensional plan that can be folded to make a
three-dimensional model of a solid *see 7·6 Surface Area*

Example:

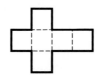

the *net* of a cube

nonagon a polygon that has nine sides

Example:

a *nonagon*

noncollinear not lying on the same straight line

noncoplanar not lying on the same plane

normal distribution represented by a bell curve, the most common distribution of most qualities across a given population *see 4·3 Analyzing Data*

Example:

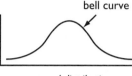

a *normal distribution*

number line a line showing numbers at regular intervals on which any real number can be indicated *see 6·5 Inequalities*

Example:

a *number line*

number symbols the symbols used in counting and measuring

Examples: $1, -\frac{1}{4}, 5, \sqrt{2}, -\pi$

number system a method of writing numbers. The Arabic *number system* is most commonly used today.

numerator the top number in a fraction. In the fraction $\frac{a}{b}$, a is the *numerator. see 2·1 Fractions and Equivalent Fractions*

obtuse angle any angle that measures more than 90° but less than 180° *see 7·1 Naming and Classifying Angles and Triangles*

Example:

an *obtuse angle*

obtuse triangle a triangle that has one obtuse angle
see 7·1 Naming and Classifying Angles and Triangles

Example:

△ABC is an *obtuse triangle*

octagon a polygon that has eight sides

Example:

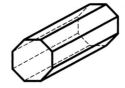

an *octagon*

octagonal prism a prism that has two octagonal bases and eight rectangular faces

Example:

an *octagonal prism*

odd numbers the set of all integers that are not multiples of 2

odds against the ratio of the number of unfavorable outcomes to the number of favorable outcomes
see 4·6 Probability

odds for the ratio of the number of favorable outcomes to the number of unfavorable outcomes
see 4·6 Probability

one-dimensional having only one measurable quality

Example: a line and a curve are *one-dimensional*

operations arithmetical actions performed on numbers, matrices, or vectors

opposite angle in a triangle, a side and an angle are said to be opposite if the side is not used to form the angle

Example:

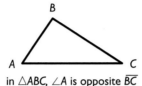

in △ABC, ∠A is opposite $\overline{BC}$

order of operations to find the answer to an equation, follow this four step process: 1) do all operations with parentheses first; 2) simplify all numbers with exponents; 3) multiply and divide in order from left to right; 4) add and subtract in order from left to right *see 1•3 Order of Operations*

ordered pair two numbers that tell the *x*-coordinate and *y*-coordinate of a point
see 6•6 Graphing on the Coordinate Plane

Example: The coordinates (3, 4) are an *ordered pair.* The *x*-coordinate is 3, and the *y*-coordinate is 4.

origin the point (0, 0) on a coordinate graph where the *x*-axis and the *y*-axis intersect

orthogonal drawing always shows three views of an object— top, side, and front. The views are drawn straight-on.

Example:

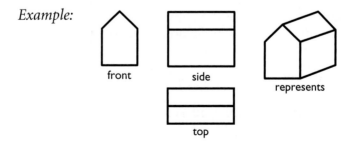

front   side

top   represents

outcome a possible result in a probability experiment

outcome grid a visual model for analyzing and representing theoretical probabilities that shows all the possible outcomes of two independent events *see 4•6 Probability*

Example:

A grid used to find the sample space for rolling a pair of dice. The outcomes are written as ordered pairs.

	1	**2**	**3**	**4**	**5**	**6**
1	(1, 1)	(2, 1)	(3, 1)	(4, 1)	(5, 1)	(6, 1)
2	(1, 2)	(2, 2)	(3, 2)	(4, 2)	(5, 2)	(6, 2)
3	(1, 3)	(2, 3)	(3, 3)	(4, 3)	(5, 3)	(6, 3)
4	(1, 4)	(2, 4)	(3, 4)	(4, 4)	(5, 4)	(6, 4)
5	(1, 5)	(2, 5)	(3, 5)	(4, 5)	(5, 5)	(6, 5)
6	(1, 6)	(2, 6)	(3, 6)	(4, 6)	(5, 6)	(6, 6)

There are 36 possible outcomes.

parallel straight lines or planes that remain a constant distance from each other and never intersect, represented by the symbol ∥

Example:

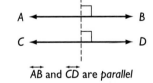

$\overleftrightarrow{AB}$ and $\overleftrightarrow{CD}$ are *parallel*

parallelogram a quadrilateral with two pairs of parallel sides *see 7•2 Naming and Classifying Polygons and Polyhedrons*

Example:

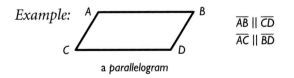

a *parallelogram*

$\overline{AB} \parallel \overline{CD}$
$\overline{AC} \parallel \overline{BD}$

parentheses the enclosing symbols ( ), which indicate that the terms within are a unit; for example, $(2 + 4) \div 2 = 3$

HOT WORDS

pattern a regular, repeating design or sequence of shapes or numbers *see Patterns, pages 61–63*

PEMDAS a reminder for the order of operations: 1) do all operations within **p**arentheses first; 2) simplify all numbers with **e**xponents; 3) **m**ultiply and **d**ivide in order from left to right; 4) **a**dd and **s**ubtract in order from left to right *see 1·3 Order of Operations*

pentagon a polygon that has five sides

Example:

a *pentagon*

percent a number expressed in relation to 100, represented by the symbol % *see 2·7 Meaning of Percent*

Example: 76 out of 100 students use computers
76 *percent* of students use computers

percent grade the ratio of the rise to the run of a hill, ramp, or incline expressed as a percent

Example:

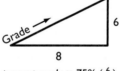

percent grade = 75% ($\frac{6}{8}$)

perfect cube a number that is the cube of an integer. For example, 27 is a *perfect cube* since $27 = 3^3$.

perfect number an integer that is equal to the sum of all its positive whole number divisors, excluding the number itself

Example: $1 \times 2 \times 3 = 6$ and $1 + 2 + 3 = 6$
6 is a *perfect number*

perfect square a number that is the square of an integer. For example, 25 is a *perfect square* since $25 = 5^2$.
see 3·2 Square Roots

perimeter the distance around the outside of a closed figure
see 7·4 Perimeter

Example:

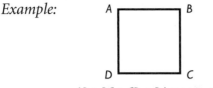

AB + BC + CD + DA = *perimeter*

permutation a possible arrangement of a group of objects.
The number of possible arrangements of n objects is
expressed by the term $n!$
see factorial, 4·5 Combinations and Permutations

perpendicular two lines or planes that intersect to form a
right angle

Example:

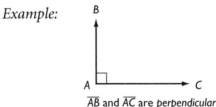

$\overline{AB}$ and $\overline{AC}$ are *perpendicular*

pi the ratio of a circle's circumference to its diameter. *Pi* is
shown by the symbol π, and is approximately equal
to 3.14. *see 7·8 Circles*

picture graph a graph that uses pictures or symbols to
represent numbers

place value the value given to a place a digit may occupy in a
numeral *see 1·1 Place Value of Whole Numbers*

place-value system a number system in which values are
given to the places digits occupy in the numeral. In the
decimal system, the value of each place is 10 times the
value of the place to its right.
see 1·1 Place Value of Whole Numbers

point one of four undefined terms in geometry used to define
all other terms. A *point* has no size.
see 6·6 Graphing on the Coordinate Plane

polygon a simple, closed plane figure, having three or more line segments as sides
see 7·2 Naming and Classifying Polygons and Polyhedrons

Examples:

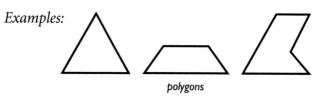

polygons

polyhedron a solid geometrical figure that has four or more plane faces
see 7·2 Naming and Classifying Polygons and Polyhedrons

Examples:

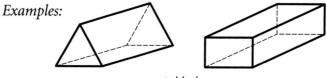

polyhedrons

population the universal set from which a sample of statistical data is selected

positive integers the set of all positive whole numbers $\{1, 2, 3, 4, 5, \ldots\}$ *see counting numbers*

positive numbers the set of all numbers that are greater than zero

Examples: $1, 1.36, \sqrt{2}, \pi$

power represented by the exponent *n*, to which a number is raised by multiplying itself *n* times
see 3·1 Powers and Exponents

Example: 7 raised to the fourth *power*
$$7^4 = 7 \times 7 \times 7 \times 7 = 2,401$$

predict to anticipate a trend by studying statistical data
see trend

price the amount of money or goods asked for or given in exchange for something else

prime factorization the expression of a composite number as a product of its prime factors
see 1·4 Factors and Multiples

Examples: $504 = 2^3 \times 3^2 \times 7$
$30 = 2 \times 3 \times 5$

prime number a whole number greater than 1 whose only factors are 1 and itself *see 1·4 Factors and Multiples*

Examples: 2, 3, 5, 7, 11

prism a solid figure that has two parallel, congruent polygonal faces (called *bases*) *see 7·2 Naming and Classifying Polygons and Polyhedrons*

Examples:

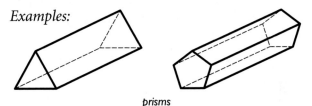

prisms

probability the study of likelihood or chance that describes the chances of an event occurring *see 4·6 Probability*

probability line a line used to order events from least likely to most likely to happen *see 4·6 Probability*

probability of events the likelihood or chance that events will occur

product the result obtained by multiplying two numbers or variables

profit the gain from a business; what is left when the cost of goods and of carrying on the business is subtracted from the amount of money taken in

project (v.) to extend a numerical model, to either greater or lesser values, in order to guess likely quantities in an unknown situation

proportion a statement that two ratios are equal
see 6•4 Ratio and Proportion

pyramid a solid geometrical figure that has a polygonal base
and triangular faces that meet at a common vertex
see 7•2 Naming and Classifying Polygons and Polyhedrons

Examples:

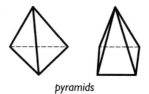

pyramids

Pythagorean Theorem a mathematical idea stating that the
sum of the squared lengths of the two shorter sides of a
right triangle is equal to the squared length of the
hypotenuse

Example:

for a right triangle, $a^2 + b^2 = c^2$

quadrant [1] one quarter of the
circumference of a circle; [2] on a
coordinate graph, one of the four regions
created by the intersection of the x-axis and
the y-axis *see 6•6 Graphing on the Coordinate Plane*

quadratic equation a polynomial equation of the second
degree, generally expressed as $ax^2 + bx + c = 0$, where
a, b, and c are real numbers and a is not equal to zero
see degree

quadrilateral a polygon that has four sides
see 7•2 Naming and Classifying Polygons and Polyhedrons

Examples:

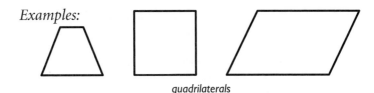

quadrilaterals

qualitative graphs a graph with words that describes such things as a general trend of profits, income, and expenses over time. It has no specific numbers.

quantitative graphs a graph that, in contrast to a qualitative graph, has specific numbers

quotient the result obtained from dividing one number or variable (the divisor) into another number or variable (the dividend)

Example:

$$24 \div 4 = 6$$

dividend | quotient

divisor

hot **words**

R

radical the indicated root of a quantity *see 3·2 Square Roots*

Examples: $\sqrt{3}$, $\sqrt[4]{14}$, $\sqrt[12]{-23}$

radical sign the root symbol $\sqrt{}$

radius a line segment from the center of a circle to any point on its circumference *see 7·8 Circles*

random sampling a population sample chosen so that each member has the same probability of being selected *see 4·1 Collecting Data*

range in statistics, the difference between the largest and smallest values in a sample *see 4·4 Statistics*

rank to order the data from a statistical sample on the basis of some criterion—for example, in ascending or descending numerical order *see 4·4 Statistics*

HOT WORDS

ranking the position on a list of data from a statistical sample based on some criterion

rate [1] fixed ratio between two things; [2] a comparison of two different kinds of units, for example, miles per hour or dollars per hour *see 6•4 Ratio and Proportion*

ratio a comparison of two numbers *see 6•4 Ratio and Proportion*

Example: the *ratio* of consonants to vowels in the alphabet is 21:5

rational numbers the set of numbers that can be written in the form $\frac{a}{b}$, where a and b are integers and b does not equal zero

Examples: $1 = \frac{1}{1}$, $\frac{2}{9}$, $3\frac{2}{7} = \frac{23}{7}$, $-.333 = -\frac{1}{3}$

ray the part of a straight line that extends infinitely in one direction from a fixed point
see 7•1 Naming and Classifying Angles and Triangles

Example:

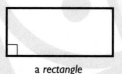

a ray

real numbers the set consisting of zero, all positive numbers, and all negative numbers. *Real numbers* include all rational and irrational numbers.

real-world data information processed by real people in everyday situations

reciprocal the result of dividing a given quantity into 1
see 2•4 Multiplication and Division of Fractions

Examples: the *reciprocal* of 2 is $\frac{1}{2}$; of $\frac{3}{4}$ is $\frac{4}{3}$; of x is $\frac{1}{x}$

rectangle a parallelogram with four right angles
see 7•2 Naming and Classifying Polygons and Polyhedrons

Example:

a rectangle

rectangular prism a prism that has rectangular bases and four rectangular faces
see 7·2 Naming and Classifying Polygons and Polyhedrons

reflection see flip, 7·3 Symmetry and Transformations

Example:

the *reflection* of a trapezoid

reflex angle any angle whose measure is greater than 180° but less than 360°

Example:

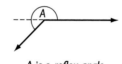

A is a *reflex angle*

regular polygon a polygon in which all sides are equal and all angles are equal

regular shape a figure in which all sides are equal and all angles are equal

relationship a connection between two or more objects, numbers, or sets. A mathematical *relationship* can be expressed in words or with numbers and letters.

repeating decimal a decimal in which a digit or a set of digits repeat infinitely

Example: 0.121212 ...

rhombus a parallelogram with all sides of equal length
see 7·2 Naming and Classifying Polygons and Polyhedrons

Example:

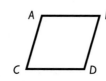

AB = CD = AC = BD

a *rhombus*

right angle an angle that measures 90°
see 7•1 Naming and Classifying Angles and Triangles

Example:

∠A is a *right angle*

right triangle a triangle with one right angle
see 7•1 Naming and Classifying Angles and Triangles

Example:

△ABC is a *right triangle*

rise the amount of vertical increase between two points

Roman numerals the numeral system consisting of the symbols I (1), V (5), X (10), L (50), C (100), D (500), and M (1,000). When a Roman symbol is preceded by a symbol of equal or greater value, the values of a symbol are added (XVI = 16). When a symbol is preceded by a symbol of lesser value, the values are subtracted (IV = 4).

root [1] the inverse of an exponent; [2] the radical sign $\sqrt{}$ indicates square root
see 3•2 Square Roots

rotation a transformation in which a figure is turned a certain number of degrees around a fixed point or line
see turn, 7•3 Symmetry and Transformations

Example:

rotation of a square

round to approximate the value of a number to a given decimal place

> *Examples:* 2.56 rounded to the nearest tenth is 2.6;
> 2.54 rounded to the nearest tenth is 2.5;
> 365 rounded to the nearest hundred is 400

row a horizontal list of numbers or terms. In spreadsheets, the labels of cells in a *row* all end with the same number (A3, B3, C3, D3 . . .) *see 9•4 Spreadsheets*

rule a statement that describes a relationship between numbers or objects

run the horizontal distance between two points

*hot*words
S

HOT WORDS

sample a finite subset of a population, used for statistical analysis
see 4•6 Probability

sampling with replacement a sample chosen so that each element has the chance of being selected more than once
see 4•6 Probability

> *Example:* A card is drawn from a deck, placed back into the deck, and a second card is drawn. Since the first card is replaced, the number of cards remains constant.

scale the ratio between the actual size of an object and a proportional representation
see 8•6 Size and Scale

scale drawing a proportionally correct drawing of an object or area at actual, enlarged, or reduced size
see 8•6 Size and Scale

scale factor the factor by which all the components of an object are multiplied in order to create a proportional enlargement or reduction
see 8•6 Size and Scale

scale size the proportional size of an enlarged or reduced representation of an object or area *see 8•6 Size and Scale*

scalene triangle a triangle with no sides of equal length

Example:

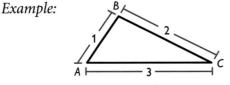

△ABC is a *scalene triangle*

scatter plot (or scatter diagram) a two-dimensional graph in which the points corresponding to two related factors (for example, smoking and life expectancy) are graphed and observed for correlation

Example:

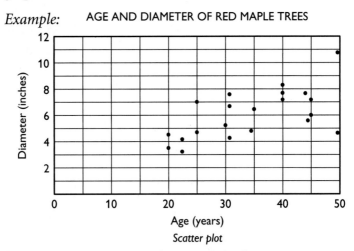

AGE AND DIAMETER OF RED MAPLE TREES

Scatter plot

scientific notation a system of writing numbers using exponents and powers of ten. A number in scientific notation is written as a number between 1 and 10 multiplied by a power of ten.

Examples: $9{,}572 = 9.572 \times 10^3$ and $0.00042 = 4.2 \times 10^{-4}$

segment two points and all the points on the line between them *see 7•1 Naming and Classifying Angles and Triangles*

sequence *see page 62*

series *see page 62*

side a line segment that forms an angle or joins the vertices of a polygon *see 7·4 Perimeter*

sighting measuring a length or angle of an inaccessible object by lining up a measuring tool with one's line of vision

signed number a number preceded by a positive or negative sign *see 1·5 Integer Operations*

significant digit the digit in a number that indicates its precise magnitude

> *Example:* 297,624 rounded to three significant digits is 298,000; 2.97624 rounded to three significant digits is 2.98

similar figures have the same shape but are not necessarily the same size *see 8·6 Size and Scale*

> *Example:*

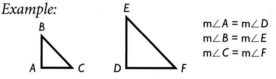

$$m\angle A = m\angle D$$
$$m\angle B = m\angle E$$
$$m\angle C = m\angle F$$

triangles ABC and DEF are *similar figures*

similarity *see similar figures*

simulation a mathematical experiment that approximates real-world process

single bar graph a way of displaying related data using one horizontal or vertical bar to represent each data item *see 4·2 Displaying Data*

slide to move a shape to another position without rotating or reflecting it
see translation, 7·3 Symmetry and Transformations

> *Example:*

the *slide* of a trapezoid

slope [1] a way of describing the steepness of a line, ramp, hill, and so on; [2] the ratio of the rise to the run

slope angle the angle that a line forms with the *x*-axis or other horizontal

slope ratio the slope of a line as a ratio of the rise to the run

solid a three-dimensional shape

solution the answer to a mathematical problem. In algebra, a *solution* usually consists of a value or set of values for a variable.

special cases a number or set of numbers, such as 0, 1, fractions and negative numbers, that is considered when determining whether or not a rule is always true

speed the rate at which an object moves

speed-time graph a graph used to chart how the speed of an object changes over time

sphere a perfectly round geometric solid, consisting of a set of points equidistant from a center point

Example:

a *sphere*

spinner a device for determining outcomes in a probability experiment

Example:

a *spinner*

spiral *see page 63*

spreadsheet a computer tool where information is arranged into cells within a grid and calculations are performed within the cells. When one cell is changed, all other cells that depend on it automatically change.
see 9·4 Spreadsheets

square a rectangle with congruent sides
see 7·2 Naming and Classifying Polygons and Polyhedrons

Example:

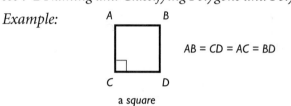

$AB = CD = AC = BD$

a *square*

square to multiply a number by itself; shown by the exponent 2
see exponent, see 3·1 Powers and Exponents

Example: $4^2 = 4 \times 4 = 16$

square centimeter a unit used to measure the size of a surface; the equivalent of a square measuring one centimeter on each side *see 8·3 Area, Volume, and Capacity*

square foot a unit used to measure the size of a surface; the equivalent of a square measuring one foot on each side *see 8·3 Area, Volume, and Capacity*

square inch a unit used to measure the size of a surface; the equivalent of a square measuring one inch on each side *see 8·3 Area, Volume, and Capacity*

square meter a unit used to measure the size of a surface; the equivalent of a square measuring one meter on each side *see 8·3 Area, Volume, and Capacity*

square number *see page 63*

Examples: 1, 4, 9, 16, 25, 36

square pyramid a pyramid with a square base

square root a number that when multiplied by itself produces a given number. For example, 3 is the *square root* of 9. *see 3·2 Square Roots*

Example: $3 \times 3 = 9$; $\sqrt{9} = 3$

HOT WORDS

square root sign the mathematical sign $\sqrt{}$; indicates that the square root of a given number is to be calculated *see 3·2 Square Roots*

standard measurement commonly used measurements, such as the meter used to measure length, the kilogram used to measure mass, and the second used to measure time *see Chapter 8 Measurement*

statistics the branch of mathematics concerning the collection and analysis of data *see 4·4 Statistics*

steepness a way of describing the amount of incline (or slope) of a ramp, hill, line, and so on

stem the ten-digit of an item of numerical data between 1 and 99 *see 4·2 Displaying Data*

stem-and-leaf plot a method of presenting numerical data between 1 and 99 by separating each number into its ten-digit (stem) and its unit-digit (leaf) and then arranging the data in ascending order of the ten-digits *see 4·2 Displaying Data*

Example:

stem	leaf
0	6
1	1 8 2 2 5
2	6 1
3	7
4	3
5	8

a *stem-and-leaf plot* for the data set 11, 26, 18, 12, 12, 15, 43, 37, 58, 6, and 21

straight angle an angle that measures 180°; a straight line

stratified random sampling a series of random samplings, each of which is taken from a specific part of the population. For example, a two-part sampling might involve taking separate samples of men and women.

strip graph a graph indicating the sequence of outcomes. A *strip graph* helps to highlight the differences among individual results and provides a strong visual representation of the concept of randomness.

Example: Outcomes of a coin toss
H = heads
T = tails

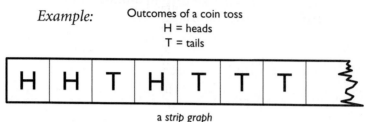

a *strip graph*

sum the result of adding two numbers or quantities

Example: 6 + 4 = 10

10 is the *sum* of the two addends, 6 and 4

surface area the sum of the areas of all the faces of a geometric solid, measured in square units
see 7•6 Surface Area

Example:

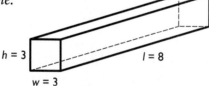

the *surface area* of this rectangular prism is
2(3 × 3) + 4(3 × 8) = 114 square units

survey a method of collecting statistical data in which people are asked to answer questions *see 4•1 Collecting Data*

symmetry *see line of symmetry*

Example:

this hexagon has *symmetry* around the dotted line

hot **words**

T

table a collection of data arranged so that information can be easily seen
see 4·2 Displaying Data

tally marks marks made for certain numbers of objects in keeping account. For example, *IIII III* = 8

term product of numbers and variables;

x, ax^2, $2x^4y^2$, and $-4ab$ are four examples of a *term*

terminating decimal a decimal with a finite number of digits

tessellation *see page 63*

Examples:

tessellations

tetrahedron a geometrical solid that has four triangular faces
see 7·2 Naming and Classifying Polygons and Polyhedrons

Example:

a *tetrahedron*

theoretical probability the ratio of the number of favorable outcomes to the total number of possible outcomes
see 4·6 Probability

three-dimensional having three measurable qualities: length, height, and width

tiling completely covering a plane with geometric shapes
see tessellation page 63

time in mathematics, the element of duration, usually represented by the variable t *see 8·5 Time*

total distance the amount of space between a starting point and an endpoint, represented by d in the equation
$d = s$ (speed) $\times t$ (time)

total distance graph a coordinate graph that shows
cumulative distance traveled as a function of time

total time the duration of an event, represented by *t* in the
equation $t = d$ (distance) $/ s$ (speed)

transformation a mathematical process that changes the
shape or position of a geometric figure *see reflection,
rotation, translation, 7•3 Symmetry and Transformations*

translation a transformation in which a geometric figure is slid
to another position without rotation or reflection
see slide, 7•3 Symmetry and Transformations

trapezoid a quadrilateral with only one pair of parallel sides
see 7•2 Naming and Classifying Polygons and Polyhedrons

Example:

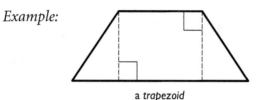

a *trapezoid*

tread the horizontal depth of one step on a stairway

tree diagram a connected, branching graph used to diagram
probabilities or factors
*see 1•4 Factors and Multiples, 4•5 Combinations and
Permutations*

Example:

a *tree diagram*

trend a consistent change over time in the statistical data
representing a particular population

triangle a polygon that has three sides
see 7•1 Naming and Classifying Angles and Triangles

HOT WORDS

triangular numbers *see page 63*

triangular prism a prism with two triangular bases and three rectangular sides *see prism*

turn to move a geometric figure by rotating it around a point *see rotation, 7·3 Symmetry and Transformations*

Example:

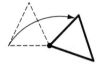

the *turning* of a triangle

two-dimensional having two measurable qualities: length and width

unequal probabilities different likelihoods of occurrence. Two events have *unequal probabilities* if one is more likely to occur than the other.

unfair where the probability of each outcome is not equal

union a set that is formed by combining the members of two or more sets, as represented by the symbol ∪. The *union* contains all members previously contained in either set *see 5·3 Sets*

Example:

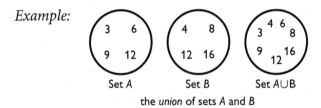

the *union* of sets A and B

unit cost the cost of an item expressed in a standard measure, such as *per ounce* or *per pint* or *each*

unit rate the rate in lowest terms

>*Example:* 120 miles in two hours is equivalent to a *unit rate* of 60 miles per hour

variable a letter or other symbol that represents a number or set of numbers in an expression or an equation *see 6•1 Setting Up Expressions and Equations*

>*Example:* in the equation $x + 2 = 7$, the variable is x

Venn diagram a pictorial means of representing the relationships between sets *see 5•3 Sets*

>*Example:*

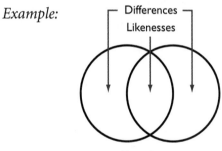

a *Venn diagram*

vertex (pl. *vertices*) the common point of two rays of an angle, two sides of a polygon, or three or more faces of a polyhedron

>*Examples:*

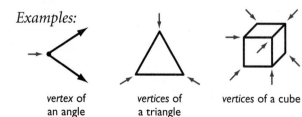

vertex of an angle   vertices of a triangle   vertices of a cube

HOT WORDS

vertex of tessellation the point where three or more
tessellating figures come together

Example:

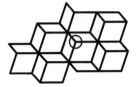

vertex of tessellation
(in the circle)

vertical a line that is perpendicular to a horizontal base line

Example:

A

base   B

$\overline{AB}$ is *vertical* to the base
of this triangle

volume the space occupied by a solid, measured in cubic units
see 7·7 Volume

Example:

h = 2

l = 5

w = 3

the *volume* of this rectangular prism is 30 cubic units
2 × 3 × 5 = 30

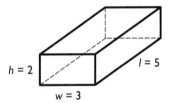

weighted average a statistical average in
which each element in the sample is
given a certain relative importance, or
weight. For example, to find the accurate
average percentage of people who own cars in three
towns with different-sized populations, the largest
town's percentage would have to be *weighted*.

what-if questions a question posed to frame, guide, and extend a problem

whole numbers the set of all counting numbers plus zero

Examples: 0, 1, 2, 3, 4, 5

width a measure of the distance of an object from side to side

x-axis the horizontal reference line in the coordinate graph *see 6·6 Graphing on the Coordinate Plane*

x-intercept the point at which a line or curve cuts across the x-axis

y-axis the vertical reference line in the coordinate graph *see 6·6 Graphing on the Coordinate Plane*

y-intercept the point at which a line or curve cuts across the y-axis *see 6·8 Slope and Intercept*

zero-pair one positive cube and one negative cube used to model signed number arithmetic

HOT WORDS

Formulas

Area (*see 7•5*)

circle	$A = \pi r^2$ (pi $\times$ square of the radius)
parallelogram	$A = bh$ (base $\times$ height)
rectangle	$A = lw$ (length $\times$ width)
square	$A = s^2$ (side squared)
trapezoid	$A = \frac{1}{2}h(b_1 + b_2)$
	($\frac{1}{2} \times$ height $\times$ sum of the bases)
triangle	$A = \frac{1}{2}bh$ ($\frac{1}{2} \times$ base $\times$ height)

Volume (*see 7•7*)

cone	$V = \frac{1}{3}\pi r^2 h$
	($\frac{1}{3} \times$ pi $\times$ square of the radius $\times$ height)
cylinder	$V = \pi r^2 h$
	(pi $\times$ square of the radius $\times$ height)
prism	$V = Bh$ (area of the base $\times$ height)
pyramid	$V = \frac{1}{3}Bh$ ($\frac{1}{3} \times$ area of the base $\times$ height)
rectangular prism	$V = lwh$ (length $\times$ width $\times$ height)
sphere	$V = \frac{4}{3}\pi r^3$ ($\frac{4}{3} \times$ pi $\times$ cube of the radius)

Perimeter (*see 7•4*)

parallelogram	$P = 2a + 2b$ (2 $\times$ side a + 2 $\times$ side b)
rectangle	$P = 2l + 2w$ (twice length + twice width)
square	$P = 4s$ (4 $\times$ side)
triangle	$P = a + b + c$ (side a + side b + side c)

Circumference (*see 7•8*)

circle	$C = \pi d$ (pi $\times$ diameter)
	or
	$C = 2\pi r$ (2 $\times$ pi $\times$ radius)

Formulas

Probability (see 4•6)

The *Experimental Probability* of an event is equal to the total number of times a favorable outcome occurred, divided by the total number of times the experiment was done.

$$\frac{Experimental}{Probability} = \frac{favorable\ outcomes\ that\ occurred}{total\ number\ of\ experiments}$$

The *Theoretical Probability* of an event is equal to the number of favorable outcomes, divided by the total number of possible outcomes.

$$\frac{Theoretical}{Probability} = \frac{favorable\ outcomes}{possible\ outcomes}$$

Other

Distance $d = rt$ (rate $\times$ time)

Interest $i = prt$ (principle $\times$ rate $\times$ time)

PIE   Profit = Income $-$ Expenses

Symbols

{ }	set	$\overline{AB}$	segment AB
Ø	the empty set	$\overrightarrow{AB}$	ray AB
⊆	is a subset of	$\overleftrightarrow{AB}$	line AB
∪	union	$\triangle ABC$	triangle ABC
∩	intersection	$\angle ABC$	angle ABC
>	is greater than	$m\angle ABC$	measure of angle ABC
<	is less than		
≥	is greater than or equal to	AB or $m\overline{AB}$	length of segment AB
≤	is less than or equal to	$\overset{\frown}{AB}$	arc AB
=	is equal to		
≠	is not equal to	!	factorial
°	degree	$_nP_r$	permutations of n things taken r at a time
%	percent		
$f(n)$	function, f of n	$_nC_r$	combinations of n things taken r at a time
$a{:}b$	ratio of a to b, $\frac{a}{b}$		
$\lvert a \rvert$	absolute value of a	$\sqrt{}$	square root
$P(E)$	probability of an event E	$\sqrt[3]{}$	cube root
π	pi	'	foot
⊥	is perpendicular to	"	inch
‖	is parallel to	÷	divide
≅	is congruent to	/	divide
~	is similar to	*	multiply
≈	is approximately equal to	×	multiply
∠	angle	·	multiply
∟	right angle	+	add
△	triangle	−	subtract

SYMBOLS

Patterns

arithmetic sequence a sequence of numbers or terms that have a common difference between any one term and the next in the sequence. In the following sequence, the common difference is seven, so $8 - 1 = 7$; $15 - 8 = 7$; $22 - 15 = 7$, and so forth.

Example: 1, 8, 15, 22, 29, 36, 43, . . .

Fibonacci numbers a sequence in which each number is the sum of its two predecessors. Can be expressed as $x_n = x_{n-2} + x_{n-1}$. The sequence begins: 1, 1, 2, 3, 5, 8, 13, 21, 34, 55, . . .

Example:

1, 1, 2, 3, 5, 8, 13, 21, 34, 55 ...
1+1=2
1+2=3
2+3=5
3+5=8

geometric sequence a sequence of terms in which each term is a constant multiple, called the *common ratio,* of the one preceding it. For instance, in nature, the reproduction of many single-celled organisms is represented by a progression of cells splitting in two in a growth progression of 1, 2, 4, 8, 16, 32, . . ., which is a geometric sequence in which the common ratio is 2.

harmonic sequence a progression $a_1, a_2, a_3, \ldots$ for which the reciprocals of the terms, $\frac{1}{a_1}, \frac{1}{a_2}, \frac{1}{a_3}, \ldots$ form an arithmetic sequence. For instance, in most musical tones, the frequencies of the sound waves are integer multiples of the fundamental frequency.

Lucas numbers a sequence in which each number is the sum of its two predecessors. Can be expressed as $x_n = x_{n-2} + x_{n-1}$

The sequence begins: 1, 3, 4, 7, 11, 18, 29, 47, . . .

magic square a square array of different numbers in which rows, columns, and diagonals add up to the same total

Example:

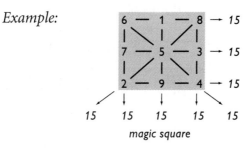

magic square

Pascal's triangle a triangular arrangement of numbers. Blaise Pascal (1623–1662) developed techniques for applying this arithmetic triangle to probability patterns.

Example:

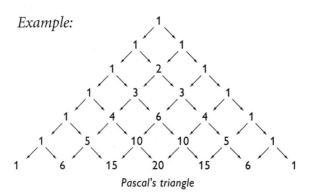

Pascal's triangle

sequence a set of elements, especially numbers, arranged in order according to some rule

series the sum of the terms of a sequence

spiral a plane curve traced by a point moving around a fixed point while continuously increasing or decreasing its distance from it

Example:

The shape of a chambered nautilus shell is a *spiral.*

square numbers a sequence of numbers that can be shown by dots arranged in the shape of a square. Can be expressed as x^2. The sequence begins 1, 4, 9, 16, 25, 36, 49, . . .

Example:

. : ::: :::: ::::: ::::::
1   4   9   16   25   36

square numbers

tessellation a tiling pattern made of repeating polygons that fills a plane completely, leaving no gaps

Example:

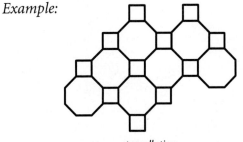

tessellation

triangular numbers a sequence of numbers that can be shown by dots arranged in the shape of a triangle. Any number in the sequence can be expressed as $x_n = x_{n-1} + n$. The sequence begins 1, 3, 6, 10, 15, 21, . . .

Example:

triangular numbers

PATTERNS

PART TWO

hot topics

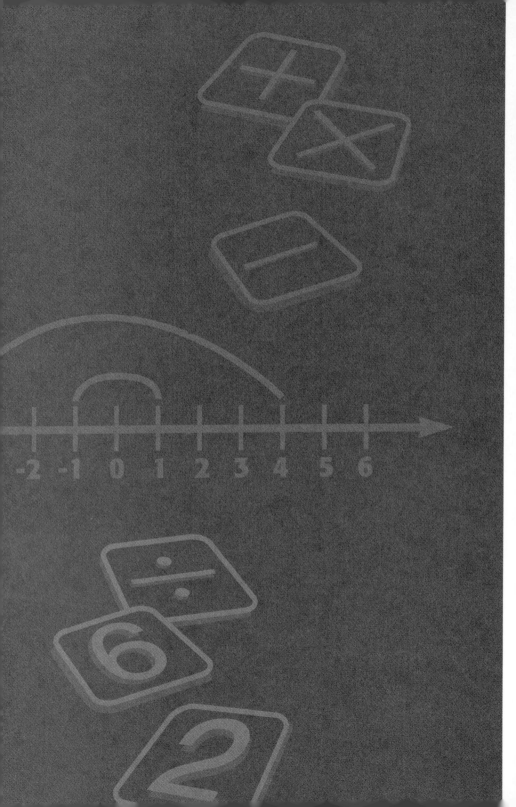

Numbers and Computation

What do you already know?

You can use the problems and the list of words that follow to see what you already know about this chapter. The answers to the problems are in Hot Solutions at the back of the book, and the definitions of the words are in Hot Words at the front of the book. You can find out more about a particular problem or word by referring to the boldfaced topic number (for example, **1•2**).

Problem Set

Give the value of the 3 in each number. **1•1**
1. 237,514
2. 736,154,987

3. Write 24,378, using expanded notation. **1•1**
4. Write in order from greatest to least: 56,418; 566,418; 5,618; 496,418 **1•1**
5. Round 52,564,764 to the nearest ten, thousand, and million. **1•1**

Solve. **1•2**
6. 258×0
7. $(5 \times 3) \times 1$
8. $3,589 + 0$
9. 0×1

Solve. Use mental math if you can. **1•2**
10. $4 \times (31 + 69)$
11. $25 \times 16 \times 4$

Use parentheses to make each expression true. **1•3**
12. $4 + 6 \times 5 = 50$
13. $10 + 14 \div 3 + 3 = 4$

Is it a prime number? Write Yes or No. **1•4**
14. 99
15. 105
16. 106
17. 97

Write the prime factorization for each. **1•4**
18. 33
19. 105
20. 180

Find the GCF for each pair. **1•4**
21. 15 and 30
22. 14 and 21
23. 18 and 120

Find the LCM for each pair. **1•4**
24. 3 and 15
25. 12 and 8
26. 16 and 40

27. What is the least common multiple of 2, 3, and 16? **1•4**

Give the absolute value of the integer. Then write its opposite. **1•5**

28. -6 29. 13
30. -15 31. 25

Add or subtract. **1•5**

32. $9 + (-3)$ 33. $4 - 5$
34. $-9 + (-9)$ 35. $3 - (-3)$
36. $-8 - (-8)$ 37. $-6 + 8$

Compute. **1•5**

38. $-4 \times (-7)$ 39. $48 \div (-12)$
40. $-42 \div (-6)$ 41. $(-4 \times 3) \times (-3)$
42. $3 \times [-6 + (-4)]$ 43. $-5 [5 - (-7)]$

44. What can you say about the product of a negative integer and a positive integer? **1•5**
45. What can you say about the sum of two positive integers? **1•5**

CHAPTER 1

hot **words**

absolute value **1•5**
approximation **1•1**
associative property **1•2**
common factor **1•4**
commutative property **1•2**
composite number **1•4**
distributive property **1•2**
expanded notation **1•1**
factor **1•4**
greatest common factor **1•4**

least common multiple **1•4**
multiple **1•4**
negative integer **1•5**
negative number **1•5**
number system **1•1**
operation **1•3**
PEMDAS **1•3**
place value **1•1**
positive integer **1•5**
prime factorization **1•4**
prime number **1•4**
round **1•1**

1·1 Place Value of Whole Numbers

Understanding Our Number System

You know that our **number system** is based on 10 and that the value of each place is 10 times the value of the place to its right. The value of a digit is the product of that digit and its **place value.** For instance, in the number 5,700, the 5 has a value of five thousands and the 7 has a value of seven hundreds.

A *place-value* chart can help you read numbers. In the chart, each group of three digits is called a *period.* Commas separate the periods. The chart below shows the area of Asia, the largest continent. The area is about 17,300,000 square miles. That is nearly twice the size of North America.

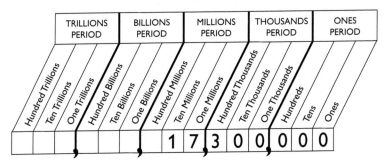

To read a large number, think of the periods. At each comma, say the name of the period.

17,300,000 reads: seventeen million three hundred thousand.

Check It Out

Give the value of the 3 in each number.
1. 14,038
2. 843,000,297

Write each number in words.
3. 40,306,200
4. 14,030,500,000,000

Using Expanded Notation

To show the place values of the digits in a number, you can write the number using **expanded notation.**

You can write 50,203 using expanded notation.

$$50,203 = 50,000 + 200 + 3$$

- Write the ten thousands. $(5 \times 10,000)$
- Write the thousands. $(0 \times 1,000)$
- Write the hundreds. (2×100)
- Write the tens. (0×10)
- Write the ones. (3×1)

So $50,203 = (5 \times 10,000) + (2 \times 100) + (3 \times 1)$.

Check It Out

Use expanded notation to write each number.

5. 83,046   6. 300,285

Comparing and Ordering Numbers

When you compare numbers, there are exactly three possibilities: the first number is greater than the second $(2 > 1)$; the second is greater than the first $(3 < 4)$; or the two numbers are equal $(6 = 6)$. When ordering several numbers, compare the numbers two at a time.

COMPARING NUMBERS

Compare 35,394 and 32,915.

- Line up the digits, starting with the ones.

  35,394

  32,915

- Beginning at the left, look at the digits in order. Find the first place where they differ.

  The digits in the thousands place differ.

- The number with the greater digit is the greater.

$5 > 2$. So 35,394 is greater than 32,915.

PLACE VALUE

Check It Out

Write >, <, or =.

7. 228,497 ☐ 238,006   8. 52,004 ☐ 51,888

Write in order from least to greatest.

9. 56,302; 52,617; 6,520; 526,000

Using Approximations

For many situations, using an **approximation** makes sense. For instance, it is reasonable to use a rounded number to express population. You might say that the population of a place is "about 60,000" rather than saying it's "58,889."

Use this rule to **round** numbers. Look at the digit to the right of the place to which you are rounding. If the digit is 5 or greater, round up. If it is less than 5, round down.

Round 123,456 to the nearest hundred.

$$\text{Hundreds}$$
$$1\,23,456$$
$$\uparrow$$
$$5 \geq 5$$

So 123,456 rounds to 123,500.

Check It Out

10. Round 32,438 to the nearest hundred.
11. Round 558,925 to the nearest ten thousand.
12. Round 2,479,500 to the nearest million.
13. Round 369,635 to the nearest hundred thousand.

 **EXERCISES**

Give the value of the 4 in each number.
1. 481,066
2. 628,014,257

Write each number in words.
3. 22,607,400
4. 3,040,680,000,000

Use expanded notation to write each number.
5. 46,056
6. 4,800,325

Write >, <, or =.
7. 436,252 ☐ 438,352
8. 85,106 ☐ 58,995

Write in order from least to greatest.
9. 38,388; 83,725; 18,652; 380,735

Round 48,463,522 to each place indicated.
10. nearest ten
11. nearest thousand
12. nearest hundred thousand
13. nearest ten million

Solve.
14. In the first year, a video game had total sales of $226,520,000. During the second year, sales were $239,195,200. Did the game earn more money or less money in the second year? How do you know?
15. About 2,000,000 people visited the aquarium last year. If this number was rounded to the nearest million, what was the greatest number of visitors? What was the least number?

 Properties

Commutative and Associative Properties

The operations of addition and multiplication share special properties because multiplication is repeated addition.

Both addition and multiplication are **commutative.** This means that the order doesn't change the sum or the product. If we let a and b be any whole numbers, then

$5 + 3 = 3 + 5$ and $5 \times 3 = 3 \times 5$

$a + b = b + a$ and $a \times b = b \times a$

Both addition and multiplication are **associative.** This means that grouping addends or factors will not change the sum or the product.

$(5 + 7) + 9 = 5 + (7 + 9)$ and $(3 \times 2) \times 4 = 3 \times (2 \times 4)$

$(a + b) + c = a + (b + c)$ and $(a \times b) \times c = a \times (b \times c)$

Subtraction and division do not share these properties. For example:

$6 - 3 = 3$, but $3 - 6 = -3$; therefore $6 - 3 \neq 3 - 6$

$6 \div 3 = 2$, but $3 \div 6 = 0.5$; therefore $6 \div 3 \neq 3 \div 6$

$(4 - 2) - 1 = 1$, but $4 - (2 - 1) = 3$;
therefore $(4 - 2) - 1 \neq 4 - (2 - 1)$

$(4 \div 2) \div 2 = 1$, but $4 \div (2 \div 2) = 4$;
therefore $(4 \div 2) \div 2 \neq 4 \div (2 \div 2)$

Check It Out

Write Yes or No.

1. $3 \times 7 = 7 \times 3$
2. $10 - 5 = 5 - 10$
3. $(8 \div 2) \div 2 = 8 \div (2 \div 2)$
4. $4 + (5 + 6) = (4 + 5) + 6$

Properties of One and Zero

When you add 0 to any number, the sum is that number. This is called the *zero (or identity) property of addition.*

$32 + 0 = 32$

When you multiply any number by 1, the product is that number. This is called the *one (or identity) property of multiplication.*

$32 \times 1 = 32$

But the product of any number and 0 is 0. This is called the *zero property of multiplication.*

$32 \times 0 = 0$

 Check It Out
Solve.

5. $24{,}357 \times 1$ 6. $99 + 0$
7. $6 \times (5 \times 0)$ 8. $(3 \times 0.5) \times 1$

Distributive Property

The **distributive property** is important because it combines both addition and multiplication. This property states that multiplying a sum by a number is the same as multiplying each addend by that number and then adding the two products.

$3(8 + 2) = (3 \times 8) + (3 \times 2)$

If we let *a, b,* and *c* be any whole numbers, then

$a \times (b + c) = (a \times b) + (a \times c)$

 Check It Out
Rewrite each expression using the distributive property.

9. $3 \times (3 + 6)$
10. $(5 \times 8) + (5 \times 7)$

Shortcuts for Adding and Multiplying

Use the properties to help you perform some computations mentally.

$$77 + 56 + 23 = (77 + 23) + 56 = 100 + 56 = 156$$

↑

Use commutative
and associative properties.

↓

$$4 \times 9 \times 25 = (4 \times 25) \times 9 = 100 \times 9 = 900$$

$$8 \times 340 = (8 \times 300) + (8 \times 40) = 2,400 + 320 = 2,720$$

↑

Use distributive
property.

Number Palindromes

Do you notice anything unusual about this word, name, or sentence?

noon   Otto
Was it a can on a cat I saw?

Each one is a *palindrome*—a word, name, or sentence that reads the same forwards and backwards. It is easy to make up number palindromes using three or more digits, like 323 or 7227. But it is harder to make up a number sentence that is the same when you read its digits from either direction, such as 10989 x 9 = 98901. Try it and see!

 **EXERCISES**

Write Yes or No.
 1. $7 \times 21 = 21 \times 7$
 2. $3 \times 4 \times 7 = 3 \times 7 \times 4$
 3. $3 \times 140 = (3 \times 100) \times (3 \times 40)$
 4. $b \times (p + r) = bp + br$
 5. $(2 \times 3 \times 5) = (2 \times 3) + (2 \times 5)$
 6. $a \times (c + d + e) = ac + ad + ae$
 7. $11 - 6 = 6 - 11$
 8. $12 \div 3 = 3 \div 12$

Solve.
 9. $22{,}350 \times 1$
10. $278 + 0$
11. $4 \times (0 \times 5)$
12. $0 \times 3 \times 15$
13. 0×1
14. $2.8 + 0$
15. 4.25×1
16. $(3 + 6 + 5) \times 1$

Rewrite each expression using the distributive property.
17. $5 \times (8 + 4)$
18. $(8 \times 12) + (8 \times 8)$
19. 4×350

Solve. Use mental math if you can.
20. $5 \times (27 + 3)$
21. $6 \times (21 + 79)$
22. 7×220
23. 25×8
24. $2 + 63 + 98$
25. $150 + 50 + 450$
26. 130×6
27. $12 \times 50 \times 2$

28. Give an example to show that subtraction is not associative.
29. Give an example to show that division is not commutative.
30. Give an example to show the zero (or identity) property of addition.

1·3 Order of Operations

Understanding the Order of Operations

Solving a problem may involve using more than one **operation.**
Your answer can depend on the order in which you do those
operations.

For instance, take the expression $2 + 3 \times 4$.

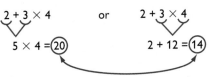

The order in which you perform
operations really makes a difference.

To make sure that there is just one answer to a series of
computations, mathematicians have agreed upon an order in
which to do the operations.

USING THE ORDER OF OPERATIONS

Simplify $2 + 8 \times (9 - 5)$.

- Simplify within the parentheses. Evaluate the exponent
  (if any).

  $$2 + 8 \times (9 - 5) = 2 + 8 \times 4$$

- Multiply or divide from left to right.

  $$2 + 8 \times 4 = 2 + 32$$

- Add or subtract from left to right.

  $$2 + 32 = 34$$

So $2 + 8 \times (9 - 5) = 34$.

 Check It Out

Simplify.

1. $20 - 2 \times 5$ 2. $3 \times (2 + 16)$

1·3 EXERCISES

Is each expression true? Write Yes or No.

1. $6 \times 3 + 4 = 22$
2. $3 + 6 \times 5 = 45$
3. $4 \times (6 + 4 \div 2) = 20$
4. $25 - (12 \times 1) = 13$
5. $(1 + 5) \times (1 + 5) = 36$
6. $(4 + 3 \times 2) + 6 = 20$
7. $35 - 5 \times 5 = 10$
8. $(9 \div 3) \times 9 = 27$

Simplify.

9. $24 - (3 \times 6)$
10. $3 \times (4 + 16)$
11. $2 \times 2 \times (8 - 5)$
12. $9 + (5 - 3)$
13. $(12 - 9) \times 5$
14. $10 + 9 \times 4$
15. $(4 + 5) \times 9$
16. $36 \div (12 + 6)$
17. $32 - (10 - 5)$
18. $24 + 6 \times (16 \div 2)$

Use parentheses to make the expression true.

19. $4 + 5 \times 6 = 54$
20. $4 \times 25 + 25 = 200$
21. $24 \div 6 + 2 = 3$
22. $10 + 20 \div 4 - 5 = 10$
23. $8 + 3 \times 3 = 17$
24. $16 - 10 \div 2 \times 4 = 44$

25. Use each number 2, 3, and 4 once to make an expression equal to 14.

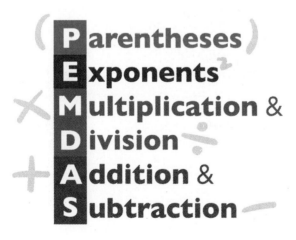

1•4 Factors and Multiples

Factors

Suppose that you want to arrange 15 small squares into a rectangular pattern.

$1 \times 15 = 15$

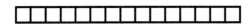

$3 \times 5 = 15$

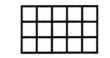

Two numbers multiplied by each other to produce 15 are considered **factors** of 15. So the factors of 15 are 1, 3, 5, and 15.

To decide whether one number is a factor of another, divide. If there is a remainder of 0, the number is a factor.

FINDING THE FACTORS OF A NUMBER

What are the factors of 20?

- Find all pairs of numbers that multiply to give the product.

  $1 \times 20 = 20 \qquad 2 \times 10 = 20 \qquad 4 \times 5 = 20$

- List the factors in order, starting with 1.

The factors of 20 are 1, 2, 4, 5, 10, and 20.

Check It Out

Find the factors of each number.

1. 6   2. 18

I·4 FACTORS AND MULTIPLES

Common Factors

Factors that are the same for two or more numbers are called **common factors.**

FINDING COMMON FACTORS

What numbers are factors of both 8 and 20?

- List the factors of the first number.

  1, 2, 4, 8

- List the factors of the second number.

  1, 2, 4, 5, 10, 20

- Common factors are the numbers that are in both lists.

  1, 2, 4

The common factors of 8 and 20 are 1, 2, and 4.

Check It Out
List the common factors of each set of numbers.
3. 8 and 12   4. 10, 15, and 20

Greatest Common Factor

The **greatest common factor** (GCF) of two whole numbers is the greatest number that is a factor of both the numbers.

One way to find the GCF is to follow these steps.
- Find the common factors.
- Choose the greatest commom factor.
What is the GCF of 12 and 40?
- Factors of 12 are 1, 2, 3, 4, 6, 12.
- Factors of 40 are 1, 2, 4, 5, 8, 10, 20, 40.
- Common factors that are in both lists are 1, 2, 4.
The greatest common factor of 12 and 40 is 4.

Check It Out
Find the GCF for each pair.
5. 8 and 10   6. 10 and 40

Divisibility Rules

Sometimes you want to know if a number is a factor of a much larger number. For instance, if you want to form teams of 3 from a group of 147 basketball players entered in a tournament, you will need to know whether 3 is a factor of 147.

You can quickly figure out whether 147 is divisible by 3 if you know the divisibility rule for 3. A number is divisible by 3 if the sum of the digits is divisible by 3. For example, 147 is divisible by 3 because $1 + 4 + 7 = 12$, and 12 is divisible by 3.

It can be helpful to know other divisibility rules. A number is divisible by:

2, if the ones digit is an even number.

3, if the sum of the digits is divisible by 3.

4, if the number formed by the last two digits is divisible by 4.

5, if the ones digit is 0 or 5.

6, if the number is divisible by 2 and 3.

8, if the number formed by the last three digits is divisible by 8.

9, if the sum of the digits is divisible by 9.

And...

Any number is divisible by **10,** if the ones digit is 0.

Check It Out

7. Is 416 divisible by 4?
8. Is 129 divisible by 9?
9. Is 462 divisible by 6?
10. Is 1,260 divisible by 5?

Prime and Composite Numbers

A **prime number** is a whole number greater than one with exactly two factors, itself and 1. Here are the first 10 prime numbers:

2, 3, 5, 7, 11, 13, 17, 19, 23, 29

Twin primes are pairs of primes whose difference is 2. (3, 5), (5, 7), and (11, 13) are examples of twin primes.

A number with more than two factors is called a **composite number.** When two composite numbers have no common factors (other than 1), they are said to be *relatively prime.* The numbers 8 and 25 are relatively prime.

One way to find out whether a number is prime or composite is to use the "sieve of Eratosthenes." Here is how it works.

• Use a chart of numbers listed in order. First skip the number 1, because it is neither prime nor composite.
• Circle the number 2 and cross out every multiple of 2.
• Next circle the number 3 and cross out every multiple of 3.
• Then continue this procedure with 5, 7, 11, and with each succeeding number that has not been crossed out.
• The prime numbers are all the circled ones. The crossed-out numbers are the composite numbers.

1 ②③ 4̸ ⑤ 6̸ ⑦ 8̸ 9̸ 1̸0̸
⑪ 1̸2̸ ⑬ 1̸4̸ 1̸5̸ 1̸6̸ ⑰ 1̸8̸ ⑲ 2̸0̸
2̸1̸ 2̸2̸ ㉓ 2̸4̸ 2̸5̸ 2̸6̸ 2̸7̸ 2̸8̸ ㉙ 3̸0̸
㉛ 3̸2̸ 3̸3̸ 3̸4̸ 3̸5̸ 3̸6̸ ㊲ 3̸8̸ 3̸9̸ 4̸0̸
㊷ 4̸2̸ ㊸ 4̸4̸ 4̸5̸ 4̸6̸ ㊼ 4̸8̸ 4̸9̸ 5̸0̸
5̸1̸ 5̸2̸ ㊾ 5̸4̸ 5̸5̸ 5̸6̸ 5̸7̸ 5̸8̸ ㊾ 6̸0̸...

Check It Out

Is it a prime number? You can use the sieve of Eratosthenes method to decide.

11. 61   12. 93

13. 83   14. 183

Prime Factorization

Every composite number can be expressed as a product of prime factors. Use a factor tree to find the prime factors.

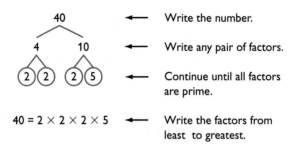

Write the number.

Write any pair of factors.

Continue until all factors are prime.

$40 = 2 \times 2 \times 2 \times 5$ ◄—— Write the factors from least to greatest.

Although the order of the factors may be different because you can start with different pairs of factors, every factor tree for 40 has the same **prime factorization.** You can also write the prime factorization using exponents.

$$40 = 2 \times 2 \times 2 \times 5 = 2^3 \times 5$$

Check It Out

What is the prime factorization of each?

15. 30

16. 80

17. 120

18. 110

Shortcut to Finding GCF

Use prime factorization to find the greatest common factor.

USING PRIME FACTORIZATION TO FIND THE GCF

Find the greatest common factor of 12 and 20.

- Find the prime factors of each number. Use a factor tree if it helps you.

  $12 = 2 \times 2 \times 3$

  $20 = 2 \times 2 \times 5$

- Find the prime factors common to both numbers.

  2 and 2

- Find their product.

  $2 \times 2 = 4$

The GCF of 12 and 20 is 2×2, or 4.

Check It Out

Use prime factorization to find the GCF of each pair of numbers.
19. 6 and 15
20. 10 and 30
21. 12 and 30
22. 24 and 36

Multiples and Least Common Multiples

The **multiples** of a number are the whole-number products when that number is a factor. In other words, you can find a multiple of a number by multiplying it by $-3, -2, -1, 0, 1, 2, 3$, and so on.

The **least common multiple** (LCM) of two numbers is the smallest nonzero number that is a multiple of both.

One way to find the LCM of a pair of numbers is to first list multiples of each and then identify the smallest one common to both. For instance, to find the LCM of 6 and 8:
• List multiples of 6: 6, 12, 18, 24, 30, ...
• List multiples of 8: 8, 16, 24, 32, 40, ...
• LCM = 24
Another way to find the LCM is to use prime factorization.

USING PRIME FACTORIZATION TO FIND THE LCM

Find the least common multiple of 6 and 8.
• Find the prime factors of each number.

$$6 = 2 \times 3 \qquad 8 = 2 \times 2 \times 2$$

• Multiply the prime factors of the least number by the prime factors of the greater number that are not factors of the least number.

$$2 \times 2 \times 2 \times 3 = 24$$

The least common multiple of 6 and 8 is 24.

Check It Out
Use either method to find the LCM.
23. 6 and 9
24. 10 and 25
25. 8 and 14
26. 15 and 50

Find the factors of each number.
1. 9
2. 24
3. 30
4. 48

Is it a prime number? Write Yes or No.
5. 51
6. 79
7. 103
8. 219

Write the prime factorization for each.
9. 55
10. 100
11. 140
12. 200

Find the GCF for each pair.
13. 8 and 24
14. 9 and 30
15. 18 and 25
16. 20 and 25
17. 16 and 30
18. 15 and 40

Find the LCM for each pair.
19. 6 and 7
20. 12 and 24
21. 16 and 24
22. 10 and 35

23. What is the divisibility rule for 6? Is 4,124 divisible by 8?
24. How do you use prime factorization to find the GCF of two numbers?
25. What is the least common multiple of 3, 4, and 5?

1·5 Integer Operations

Positive and Negative Integers

A glance through any newspaper shows that many quantities are expressed using **negative numbers.** For example, negative numbers show below-zero temperatures.

Whole numbers greater than zero are called **positive integers.** Whole numbers less than zero are called **negative integers.**

Here is the set of all integers:
    ..., −5, −4, −3, −2, −1, 0, 1, 2, 3, 4, 5, ...
The integer 0 is neither positive nor negative.

> **Check It Out**
> Write an integer to describe the situation.
> 1. 3 below zero   2. a gain of $250

Opposites of Integers and Absolute Value

Integers can describe opposite ideas. Each integer has an opposite.
    The opposite of a gain of 5 pounds is a loss of 5 pounds.
    The opposite of +5 is −5.
    The opposite of spending $3 is earning $3.
    The opposite of −3 is +3.
The **absolute value** of an integer is its distance from 0 on the number line. You write the absolute value of −2 as |−2|.

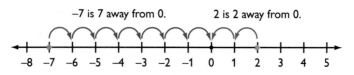

−7 is 7 away from 0.   2 is 2 away from 0.

The absolute value of 2 is 2. You write |2| = 2.
The absolute value of −7 is 7. You write |−7| = 7.

Check It Out

Give the absolute value of the integer. Then write the opposite of the original integer.

3. -12 4. $+4$

5. -8 6. 0

Adding and Subtracting Integers

Use a number line to model adding and subtracting integers.

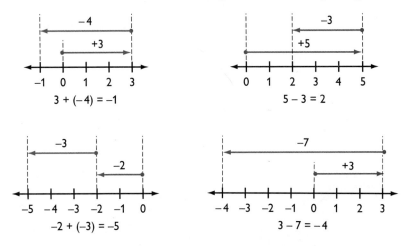

$$3 + (-4) = -1$$

$$5 - 3 = 2$$

$$-2 + (-3) = -5$$

$$3 - 7 = -4$$

Rules for Adding or Subtracting Integers		
To	Solve	Example
Add same sign	Add absolute values. Use original sign in answer.	$-3 + (-3)$: $\|-3\| + \|-3\| = 3 + 3 = 6$ So $-3 + (-3) = -6$.
Add different signs	Subtract absolute values. Use sign of addend with greatest absolute value in answer.	$-6 + 4$: $\|-6\| - \|4\| = 6 - 4 = 2$ $\|-6\| > \|4\|$ So $-6 + 4 = -2$.
Subtract	Add opposite.	$-4 - 2 = -4 + (-2) = -6$

Check It Out

7. $5 - 7$ 8. $6 + (-6)$

9. $-5 - (-7)$ 10. $0 + (-3)$

Multiplying and Dividing Integers

Multiply and divide integers as you would whole numbers. Then use these rules for writing the sign of the answer.

The product of two integers with like signs is positive. So is the quotient.

$$2 \times 3 = 6 \qquad -4 \times (-3) = 12 \qquad -12 \div (-4) = 3$$

When the signs of the two integers are different, the product is negative. So is the quotient.

$$-6 \div 3 = -2 \qquad -3 \times 5 = -15 \qquad -4 \times 10 = -40$$

Check It Out

Find the product or quotient.

11. $-3 \times (-2)$ 12. $12 \div (-4)$

13. $-15 \div (-3)$ 14. -6×9

?

Oops!

Seventeen-year-old Colin Rizzio took the **SAT** test and found a mistake in its math portion. One of the questions used the letter a to represent a number. The test makers assumed a was a positive number. But Colin Rizzio thought it could stand for any integer. Rizzio was right!

He notified the test makers by email. They had to change the test scores of 45,000 students.

Explain how $2 + a > 2$ changes if a can be positive, negative, or zero. See Hot Solutions for answer.

1·5 EXERCISES

Give the absolute value of the integer. Then write its opposite.

1. -11 2. 5
3. -5 4. 2

Add or subtract.

5. $4 - 3$ 6. $4 + (-6)$
7. $-5 - (-4)$ 8. $0 + (-3)$
9. $-2 + 6$ 10. $0 - 8$
11. $0 - (-6)$ 12. $-3 - 8$
13. $7 + (-7)$ 14. $-5 - (-8)$
15. $-2 - (-2)$ 16. $-6 + (-9)$

Find the product or quotient.

17. $-2 \times (-6)$ 18. $8 \div (-4)$
19. $-35 \div 5$ 20. -5×7
21. $4 \times (-9)$ 22. $-40 \div 8$
23. $-18 \div (-3)$ 24. $6 \times (-7)$

Compute.

25. $[-6 \times (-2)] \times 3$
26. $4 \times [2 \times (-4)]$
27. $[-3 \times (-3)] \times -3$
28. $-4 \times [3 + (-4)]$
29. $[-7 + (-3)] \times 4$
30. $-2 \times [6 - (-2)]$

31. Is the absolute value of a negative integer positive or negative?

32. If you know that the absolute value of an integer is 5, what are the possible values for that integer?

33. What can you say about the sum of two negative integers?

34. The temperature at noon was 10°F. For the next 3 hours it dropped at a rate of 3 degrees an hour. First express this change as an integer. Then give the temperature at 3:00 P.M.

35. What can you say about the product of two positive integers?

What have you learned?

You can use the problems and list of words below to see what you have learned in this chapter. You can find out more about a particular problem or word by referring to the boldfaced topic number (for example, **1•2**).

Problem Set

Give the value of the 8 in each number. **1•1**
1. 287,617
2. 758,122,907

3. Write 36,514 using expanded notation. **1•1**
4. Write in order from greatest to least: 243,254; 283,254; 83,254; and 93,254 **1•1**
5. Round 46,434,482 to the nearest ten, thousand, and million. **1•1**

Solve. **1•2**
6. 736 × 0
7. (5 × 4) × 1
8. 5,945 + 0
9. 0 × 0

Solve. Use mental math if you can. **1•2**
10. 8 × (34 + 66)
11. 50 × 15 × 2

Use parentheses to make each expression true. **1•3**
12. 5 + 7 × 2 = 24
13. 32 + 12 ÷ 4 + 5 = 40

Is it a prime number? Write Yes or No. **1•4**
14. 51
15. 102
16. 173
17. 401

Write the prime factorization for each. **1•4**
18. 35
19. 130
20. 190

Find the GCF for each pair. **1•4**
21. 16 and 36
22. 12 and 45
23. 20 and 160

Find the LCM for each pair. **1•4**
24. 5 and 10   25. 12 and 8
26. 18 and 20

27. What is the divisibility rule for 10? Is 2,550 a multiple of 10? **1•4**

Give the absolute value of the integer. Then write the opposite of the original integer. **1•5**
28. -4 29. 14
30. -17 31. -5

Add or subtract. **1•5**
32. $10 + (-9)$ 33. $3 - 8$
34. $-4 + (-4)$ 35. $2 - (-2)$
36. $-12 - (-12)$ 37. $-6 + 12$

Compute. **1•5**
38. $-9 \times (-6)$ 39. $36 \div (-12)$
40. $-54 \div (-9)$ 41. $(-4 \times 2) \times (-5)$
42. $3 \times [-6 + (-6)]$ 43. $-2 + [4 - (-9)]$

44. What can you say about the quotient of a positive integer and a negative integer? **1•5**
45. What can you say about the product of two positive integers? **1•5**

---

hot words

WRITE DEFINITIONS FOR THE FOLLOWING WORDS.

absolute value **1•5**
approximation **1•1**
associative property **1•2**
common factor **1•4**
commutative property **1•2**
composite number **1•4**
distributive property **1•2**
expanded notation **1•1**
factor **1•4**
greatest common factor **1•4**

least common multiple **1•4**
multiple **1•4**
negative integer **1•5**
negative number **1•5**
number system **1•1**
operation **1•3**
PEMDAS **1•3**
place value **1•1**
positive integer **1•5**
prime factorization **1•4**
prime number **1•4**
round **1•1**

Fractions, Decimals, and Percents

What do you already know?

You can use the problems and the list of words that follow to see what you already know about this chapter. The answers to the problems are in the Hot Solutions at the back of the book, and the definitions of the words are in Hot Words at the front of the book. To find out more about a particular problem or word, refer to the boldfaced topic number (example, 2•1).

Problem Set

1. For a family vacation, Chenelle bought 3 polo shirts for $6.75 each and 3 hats for $8.50 each. How much money did she spend? **2•6**

2. To raise money for a local charity, 28 students in the sixth grade agreed to participate in a walk-a-thon. Each of the sponsors agreed to pledge $0.45 for each mile the students walked. If the 28 students walked 20 mi each, how much money did they collect? **2•6**

3. Kelly got 4 out of 50 problems wrong on her social studies test. What percent did she get correct? **2•8**

4. A pair of shoes with a regular price of $37 is discounted 15%. What is the amount of the discount? **2•8**

5. Which fraction is not equivalent to $\frac{2}{3}$? **2•1**
   A. $\frac{4}{6}$ B. $\frac{40}{60}$ C. $\frac{12}{21}$ D. $\frac{18}{27}$

Add or subtract. Write your answers in lowest terms. **2•3**

6. $3\frac{2}{3} + \frac{1}{2}$

7. $2\frac{1}{4} - \frac{4}{5}$

8. $6 - 2\frac{1}{6}$

9. $2\frac{1}{7} + 2\frac{4}{9}$

10. Find the improper fraction and write it as a mixed number. **2•1**
    A. $\frac{4}{9}$ B. $\frac{5}{2}$ C. $3\frac{1}{2}$ D. $\frac{7}{8}$

Solve. Reduce to lowest terms. **2•4**

11. $\frac{5}{6} \times \frac{3}{8}$

12. $\frac{3}{5} \div 4\frac{1}{3}$

13. $2\frac{1}{2} \times \frac{1}{5}$

14. $3\frac{3}{7} \div 5\frac{1}{6}$

15. Give the place value of 6 in 23.064. **2•5**

16. Write in expanded form: 4.603. **2•5**

17. Write as a decimal: two hundred forty-seven thousandths. **2•5**

18. Write the following numbers in order from least to greatest: 1.655; 1.605; 16.5; 1.065. **2•5**

Find each answer as indicated. **2•6**
19. 5.466 + 12.45   20. 13.9 − 0.677
21. 4.3 × 23.67

Use a calculator to answer items 22–24. Round to the nearest tenth. **2•8**
22. What percent of 56 is 14?   23. Find 16% of 33.
24. 15 is what percent of 76?

Write each decimal as a percent. **2•9**
25. 0.68   26. 0.5

Write each fraction as a percent. **2•9**
27. $\frac{6}{100}$ 28. $\frac{56}{100}$

Write each percent as a decimal. **2•9**
29. 34%   30. 125%

Write each percent as a fraction in lowest terms. **2•9**
31. 28%   32. 130%

CHAPTER 2

hot **words**

benchmark **2•7**
common denominator **2•2**
cross product **2•1**
denominator **2•1**
discount **2•8**
equivalent **2•1**

equivalent fractions **2•1**
estimate **2•3**
factor **2•4**
fraction **2•1**
greatest common factor **2•1**
improper fraction **2•1**
least common multiple **2•2**
mixed number **2•1**

numerator **2•1**
percent **2•7**
place value **2•5**
product **2•4**
proportions **2•8**
ratio **2•7**
reciprocal **2•4**
repeating decimal **2•9**
terminating decimal **2•9**
whole number **2•1**

2·1 Fractions and Equivalent Fractions

Naming Fractions

A **fraction** can be used to name a part of a whole. The flag of Sierra Leone is divided into three equal parts: green, white, and blue. Each part, or color, of the flag represents $\frac{1}{3}$ of the whole flag. $\frac{3}{3}$ or 1 represents the whole flag.

A fraction can also name part of a set. There are four balls in the set of balls. Each ball is $\frac{1}{4}$ of the set. $\frac{4}{4}$ or 1 equals the whole set. Three of the balls are baseballs. The baseballs represent $\frac{3}{4}$ of the set. One of the four balls is a football. The football represents $\frac{1}{4}$ of the set.

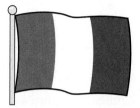

You name fractions by their **numerators** and **denominators**.

NAMING FRACTIONS

Write a fraction for the number of shaded rectangles.

- The denominator of the fraction tells the number of parts of the whole set.

    There are 5 rectangles altogether.
- The numerator of the fraction tells the number of parts under consideration.

    There are 4 shaded rectangles.
- Write the fraction:

$$\frac{\text{parts under consideration}}{\text{parts that make a whole set}} = \frac{\text{numerator}}{\text{denominator}}$$

The fraction for the number of shaded rectangles is $\frac{4}{5}$.

Check It Out
Write the fraction for each picture.

1. _____ of the circle is shaded.

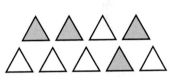

2. _____ of the triangles are shaded.

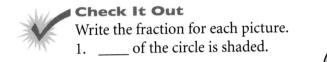

3. Draw two pictures to represent the fraction $\frac{5}{8}$. Use regions and sets.

Methods for Finding Equivalent Fractions

Equivalent fractions are fractions that describe the same amount of a region. You can use fraction pieces to show equivalent fractions.

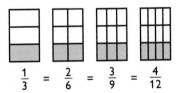

$$\frac{1}{3} = \frac{2}{6} = \frac{3}{9} = \frac{4}{12}$$

Each of the fraction pieces represents fractions equal to $\frac{1}{3}$. This makes them equivalent fractions.

2·1 EQUIVALENT FRACTIONS

Fraction Names for One

There are an infinite number of fractions that are equal to one.

Names for One				**Not Names for One**			
$\dfrac{2}{2}$	$\dfrac{365}{365}$	$\dfrac{1}{1}$	$\dfrac{5}{5}$	$\dfrac{1}{0}$	$\dfrac{3}{1}$	$\dfrac{1}{365}$	$\dfrac{11}{12}$

Since any number multiplied by one is still equal to the original number, knowing different names for one can help you find equivalent fractions.

To find a fraction that is **equivalent** to another fraction, you can multiply the original fraction by a form of one. You can also divide the numerator and denominator by the same number to get an equivalent fraction.

METHODS FOR FINDING EQUIVALENT FRACTIONS

Find a fraction equal to $\frac{9}{12}$.

• Multiply the fraction by a form of one. Or divide the numerator and denominator by the same number.

Multiply	OR	Divide
$\dfrac{9}{12} \times \dfrac{2}{2} = \dfrac{18}{24}$		$\dfrac{9 \div 3}{12 \div 3} = \dfrac{3}{4}$
$\dfrac{9}{12} = \dfrac{18}{24}$		$\dfrac{9}{12} = \dfrac{3}{4}$

Check It Out

Write two fractions equivalent to each fraction.

4. $\frac{1}{3}$ 5. $\frac{6}{12}$ 6. $\frac{3}{5}$

7. Write three fraction names for one.

Deciding if Two Fractions Are Equivalent

You can consider two fractions to be equivalent if you can show that each fraction is just a different name for the same amount.

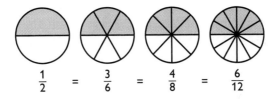

$$\frac{1}{2} \; = \; \frac{3}{6} \; = \; \frac{4}{8} \; = \; \frac{6}{12}$$

One method you can use to identify equivalent fractions is to find the **cross products** of the fractions.

DECIDING IF TWO FRACTIONS ARE EQUIVALENT

Find out whether $\frac{2}{3}$ is equivalent to $\frac{10}{15}$.

- Cross multiply the fractions.

$$\frac{2}{3} \stackrel{?}{=} \frac{10}{15}$$

$$2 \times 15 \stackrel{?}{=} 10 \times 3$$
$$30 = 30$$

- Compare the cross products.

$$30 = 30$$

- If the cross products are the same, then the fractions are equivalent.

So $\frac{2}{3} = \frac{10}{15}$.

Check It Out

Use the cross-products method to determine whether each pair of fractions is equivalent.

8. $\frac{3}{4}, \frac{27}{36}$ 9. $\frac{5}{6}, \frac{25}{30}$ 10. $\frac{15}{32}, \frac{45}{90}$

Writing Fractions in Lowest Terms

When the numerator and the denominator of a fraction have no common factor other than one, the fraction is in *lowest terms*. You can use fraction pieces to show fractions in lowest terms.

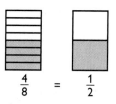

$$\frac{4}{8} = \frac{1}{2}$$

The fewest number of fraction pieces to show the equivalent of $\frac{4}{8}$ is $\frac{1}{2}$. Therefore the fraction $\frac{4}{8}$ is equal to $\frac{1}{2}$ in lowest terms.

Oseola McCarty

Miss Oseola McCarty had to leave school after sixth grade. At first, she charged $1.50 to do a bundle of laundry, later $10.00. But she always managed to save. By age 86, she'd accumulated $250,000. In 1995, she decided to donate $150,000 to endow a scholarship. Miss McCarty said, "The secret to building a fortune is compounding interest. You've got to leave your investment alone long enough for it to increase."

2·1 EQUIVALENT FRACTIONS

To express fractions in lowest terms, you can divide numerators and denominators by **greatest common factors** (GCF).

FINDING LOWEST TERMS OF FRACTIONS

Express $\frac{12}{18}$ in lowest terms.

- List the factors of the numerator.
  The factors of 12 are:

  1, 2, 3, 4, 6, 12

- List the factors of the denominator.
  The factors of 18 are:

  1, 2, 3, 6, 9, 18

- Find the greatest common factor (GCF).
  The greatest factor common to both 12 and 18 is 6.

  The GCF is 6.

- Divide the numerator and the denominator of the fraction by the GCF.

  $$\frac{12 \div 6}{18 \div 6} = \frac{2}{3}$$

- Write the fraction in lowest terms.

  $\frac{2}{3}$

$\frac{12}{18}$ expressed in lowest terms is $\frac{2}{3}$.

Check It Out

Express each fraction in lowest terms.

11. $\frac{4}{20}$

12. $\frac{9}{27}$

13. $\frac{18}{20}$

Musical Fractions

In music, notes are written on a series of lines called a *staff*. The shape of a note shows its time value—how long the note lasts when the music is played. A whole note has the longest time value. A half note is held half as long as a whole note. Other notes are held for other fractions of time compared to the whole note. Each flag makes the value of the note half of what it was before the flag was added.

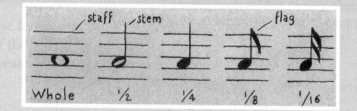

A series of short notes may be connected by a beam, instead of writing each one with a flag.

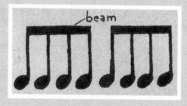

How long are each of these notes? See Hot Solutions for answer.

Write a series of notes to show that 1 whole note is equal to 2 half notes, 4 quarter notes, and 16 sixteenth notes. See Hot Solutions for answer.

Writing Improper Fractions and Mixed Numbers

You can write fractions for amounts greater than one. Fractions with a numerator greater than or equal to the denominator are called **improper fractions**.

$\frac{7}{2}$ cookies

$\frac{7}{2}$ is an improper fraction.

A **whole number** and a fraction make up a **mixed number**.

$3\frac{1}{2}$ cookies

$3\frac{1}{2}$ is a mixed number.

You can write any mixed number as an improper fraction and any improper fraction as a mixed number. You can use division to change an improper fraction to a mixed number.

CHANGING AN IMPROPER FRACTION TO A MIXED NUMBER

Change $\frac{8}{3}$ to a mixed number.

- Divide the numerator by the denominator.

$$\text{Divisor} \longrightarrow 3\overline{)\begin{array}{c} 2 \longleftarrow \text{Quotient} \\ 8 \\ \underline{6} \\ 2 \longleftarrow \text{Remainder} \end{array}}$$

- Write the mixed number.

$$\text{Quotient} \longrightarrow 2\frac{2}{3} \begin{array}{l} \longleftarrow \text{Remainder} \\ \longleftarrow \text{Divisor} \end{array}$$

2·1 EQUIVALENT FRACTIONS

You can use multiplication to change a mixed number to an improper fraction. Start by renaming the whole-number part. Rename it as an improper fraction with the same denominator as the fraction part, then add the two parts.

CHANGING A MIXED NUMBER TO AN IMPROPER FRACTION

Change $2\frac{1}{2}$ to an improper fraction. Express in lowest terms.

• Multiply the whole-number part by a version of one that has the same denominator as the fraction part.

$$2 \times \frac{2}{2} = \frac{4}{2}$$

• Add the two parts (p. 112).

$$2\frac{1}{2} = \frac{4}{2} + \frac{1}{2} = \frac{5}{2}$$

So $2\frac{1}{2} = \frac{5}{2}$.

Check It Out

Write a mixed number for each improper fraction.

14. $\frac{24}{5}$

15. $\frac{13}{9}$

16. $\frac{33}{12}$

17. $\frac{29}{6}$

Write an improper fraction for each mixed number.

18. $1\frac{7}{10}$

19. $5\frac{1}{8}$

20. $6\frac{3}{5}$

21. $7\frac{3}{7}$

2·1 EXERCISES

Write the fraction for each picture.
1. ___ of the fruits are lemons.

2. ____ of the circle is red.

3. ___ of the triangles are green.

4. ___ of the balls are basketballs.

5. ___ of the balls are baseballs.

Write the fraction.
6. four ninths
7. twelve thirteenths
8. fifteen thirds
9. two halves

Write one fraction equivalent to the given fraction.
10. $\frac{2}{5}$ 11. $\frac{1}{9}$ 12. $\frac{9}{36}$ 13. $\frac{60}{70}$

Express each fraction in lowest terms.
14. $\frac{10}{24}$ 15. $\frac{16}{18}$ 16. $\frac{36}{40}$

Write each improper fraction as a mixed number.
17. $\frac{24}{7}$ 18. $\frac{32}{5}$ 19. $\frac{12}{7}$

Write each mixed number as an improper fraction.
20. $2\frac{3}{4}$ 21. $11\frac{8}{9}$ 22. $1\frac{5}{6}$
23. $3\frac{1}{3}$ 24. $4\frac{1}{4}$ 25. $8\frac{3}{7}$

2·1 EXERCISES

2·2 Comparing and Ordering Fractions

Comparing Fractions

You can use fraction pieces to compare fractions.

$$\frac{1}{2} > \frac{2}{5}$$

$$\frac{1}{4} < \frac{5}{6}$$

You can also compare fractions if you find *equivalent fractions* (p. 100) and compare numerators.

COMPARING FRACTIONS

Compare the fractions $\frac{4}{5}$ and $\frac{5}{7}$.

• Look at the denominators.

Denominators are different.

• If the denominators are different, write equivalent fractions with a **common denominator.**

35 is the least common multiple of 5 and 7. Use 35 for the common denominator.

$$\frac{4}{5} \times \frac{7}{7} = \frac{28}{35} \qquad \frac{5}{7} \times \frac{5}{5} = \frac{25}{35}$$

• Compare the numerators.

28 > 25

• The fractions compare as the numerators compare.

$\frac{28}{35} > \frac{25}{35}$, so $\frac{4}{5} > \frac{5}{7}$.

Check It Out

Compare the fractions. Use $<$, $>$, or $=$ for each $\square$.

1. $\frac{3}{8} \square \frac{1}{2}$ 2. $\frac{7}{10} \square \frac{3}{4}$ 3. $\frac{7}{8} \square \frac{7}{10}$ 4. $\frac{3}{10} \square \frac{9}{30}$

Comparing Mixed Numbers

To compare *mixed numbers* (p. 105), first compare the whole numbers. Then compare the fractions, if necessary.

COMPARING MIXED NUMBERS

Compare $1\frac{2}{5}$ and $1\frac{4}{7}$.

- Be sure fractions are not improper.

  $\frac{2}{5}$ and $\frac{4}{7}$ are not improper.

- Compare the whole-number parts. If they are different, the one that is greater is the greater mixed number. If they are equal, go on.

  $1 = 1$

Compare the fraction parts by renaming them with a *common denominator* (p. 108).

  35 is the least common multiple of 5 and 7.
  Use 35 for the common denominator.

  $$\frac{2}{5} \times \frac{7}{7} = \frac{14}{35} \qquad \frac{4}{7} \times \frac{5}{5} = \frac{20}{35}$$

- Compare the fractions.

  $\frac{14}{35} < \frac{20}{35}$, so $1\frac{2}{5} < 1\frac{4}{7}$.

Check It Out

Compare each mixed number. Use $<$, $>$, or $=$ for each $\square$.

5. $1\frac{3}{4} \square 1\frac{2}{5}$

6. $2\frac{2}{9} \square 2\frac{1}{17}$

7. $5\frac{16}{19} \square 5\frac{4}{7}$

Ordering Fractions

To compare and order fractions, you can find equivalent fractions and then compare the numerators of the fractions.

ORDERING FRACTIONS WITH UNLIKE DENOMINATORS

Order the fractions $\frac{2}{5}$, $\frac{3}{4}$, and $\frac{3}{10}$ from least to greatest.

- Find the **least common multiple** (LCM) (p. 86) of $\frac{2}{5}$, $\frac{3}{4}$, and $\frac{3}{10}$.

Multiples of 4: 4, 8, 12, 16, ⓩ0, 24,…
Multiples of 5: 5, 10, 15, ⓩ0, 25,…
Multiples of 10: 10, ⓩ0, 30, 40,…
The LCM of 4, 5, and 10 is 20.

- Write equivalent fractions with the LCM as the common denominator.

$$\frac{2}{5} = \frac{2}{5} \times \frac{4}{4} = \frac{8}{20}$$

$$\frac{3}{4} = \frac{3}{4} \times \frac{5}{5} = \frac{15}{20}$$

$$\frac{3}{10} = \frac{3}{10} \times \frac{2}{2} = \frac{6}{20}$$

- The fractions compare as the numerators compare.

$$\frac{6}{20} < \frac{8}{20} < \frac{15}{20}, \text{ so } \frac{3}{10} < \frac{2}{5} < \frac{3}{4}.$$

Check It Out

Order the fractions from least to greatest.

8. $\frac{2}{4}, \frac{4}{5}, \frac{5}{8}$

9. $\frac{3}{4}, \frac{2}{3}, \frac{7}{12}$

10. $\frac{5}{6}, \frac{2}{3}, \frac{5}{8}$

Compare each fraction. Use $<$, $>$, or $=$.

1. $\frac{1}{2}, \frac{3}{7}$

2. $\frac{10}{12}, \frac{5}{6}$

3. $\frac{4}{5}, \frac{3}{4}$

4. $\frac{5}{8}, \frac{2}{3}$

5. $\frac{1}{5}, \frac{20}{100}$

6. $\frac{4}{3}, \frac{3}{4}$

Compare each mixed number. Use $<$, $>$, or $=$ for each $\square$.

7. $3\frac{3}{8} \square 3\frac{4}{7}$

8. $1\frac{2}{3} \square 1\frac{3}{5}$

9. $2\frac{3}{4} \square 2\frac{5}{6}$

10. $5\frac{4}{5} \square 5\frac{5}{8}$

11. $2\frac{1}{3} \square 2\frac{3}{9}$

12. $2\frac{3}{4} \square 1\frac{4}{5}$

Order the fractions and mixed numbers from least to greatest.

13. $\frac{4}{7}; \frac{1}{3}; \frac{9}{14}$

14. $\frac{2}{3}; \frac{5}{9}; \frac{4}{7}$

15. $\frac{3}{8}; \frac{1}{2}; \frac{5}{32}; \frac{3}{4}$

16. $\frac{2}{3}; \frac{5}{6}; \frac{5}{24}; \frac{7}{12}$

17. $2\frac{1}{3}; \frac{6}{3}; \frac{3}{4}; \frac{13}{4}$

18. $\frac{4}{5}; \frac{7}{10}; \frac{15}{5}; \frac{16}{10}$

Use the information about recess soccer goals to answer item 19.

Recess Soccer Goals

An-An	$\frac{2}{5}$
Derrick	$\frac{4}{7}$
Roberto	$\frac{5}{8}$
Gwen	$\frac{8}{10}$

numerator = goals made
denominator = goals attempted

19. Who was more accurate, Derrick or An-An?

20. The Wildcats won $\frac{3}{4}$ of their games. The Hawks won $\frac{5}{6}$ of theirs. The Bluejays won $\frac{7}{8}$ of theirs. Which team won the greatest fraction of their games? the least?

2·3 Addition and Subtraction of Fractions

Adding and Subtracting Fractions with Like Denominators

When you add or subtract fractions that have the same, or like, *denominators* (p. 98), you only add or subtract the *numerators* (p. 98). The denominator stays the same.

$$\frac{1}{3} + \frac{2}{3} = \frac{3}{3} = 1$$

ADDING AND SUBTRACTING FRACTIONS WITH LIKE DENOMINATORS

Add $\frac{1}{8} + \frac{5}{8}$. Subtract $\frac{8}{10} - \frac{3}{10}$.

- Add or subtract the numerators.

  $\frac{1}{8} + \frac{5}{8}$ $1 + 5 = 6$ $\frac{8}{10} - \frac{3}{10}$ $8 - 3 = 5$

- Write the result over the denominator.

  $\frac{1}{8} + \frac{5}{8} = \frac{6}{8}$ $\frac{8}{10} - \frac{3}{10} = \frac{5}{10}$

- Simplify, if possible.

  $\frac{6}{8} = \frac{3}{4}$ $\frac{5}{10} = \frac{1}{2}$

So $\frac{1}{8} + \frac{5}{8} = \frac{3}{4}$ and $\frac{8}{10} - \frac{3}{10} = \frac{1}{2}$.

Check It Out

Add or subtract. Express your answers in lowest terms.

1. $\frac{5}{6} + \frac{7}{6}$ 2. $\frac{4}{25} + \frac{2}{25}$

3. $\frac{11}{23} - \frac{6}{23}$ 4. $\frac{13}{16} - \frac{7}{16}$

Adding and Subtracting Fractions with Unlike Denominators

To add or subtract fractions with unlike denominators, you rename the fractions so they have the same denominator.

$$\frac{2}{3} + \frac{1}{6} = \qquad \qquad \frac{4}{6} + \frac{1}{6} = \frac{5}{6}$$

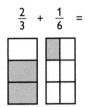 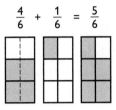

Change $\frac{2}{3}$ to sixths so the fractions have the same denominator. Then add.

To add or subtract fractions with unlike denominators, you need to change them to equivalent fractions with common, or like, denominators before you find the sum or difference.

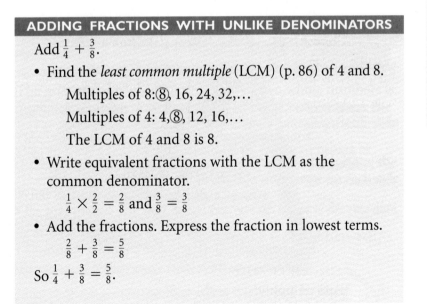

ADDING FRACTIONS WITH UNLIKE DENOMINATORS

Add $\frac{1}{4} + \frac{3}{8}$.

- Find the *least common multiple* (LCM) (p. 86) of 4 and 8.

    Multiples of 8: 8, 16, 24, 32,...

    Multiples of 4: 4, 8, 12, 16,...

    The LCM of 4 and 8 is 8.

- Write equivalent fractions with the LCM as the common denominator.

    $\frac{1}{4} \times \frac{2}{2} = \frac{2}{8}$ and $\frac{3}{8} = \frac{3}{8}$

- Add the fractions. Express the fraction in lowest terms.

    $\frac{2}{8} + \frac{3}{8} = \frac{5}{8}$

    So $\frac{1}{4} + \frac{3}{8} = \frac{5}{8}$.

Check It Out

Add or subtract. Express your answers in lowest terms.

5. $\frac{3}{4} + \frac{1}{2}$ 6. $\frac{5}{6} - \frac{2}{3}$

7. $\frac{1}{5} + \frac{1}{2}$ 8. $\frac{2}{3} - \frac{1}{12}$

Adding and Subtracting Mixed Numbers

Adding and subtracting mixed numbers is similar to adding and subtracting fractions. Sometimes you have to rename your number to subtract. Sometimes you will have an improper fraction to simplify.

Adding Mixed Numbers with Common Denominators

To add *mixed numbers* (p. 105) with common denominators, you just need to write the sum of the numerators over the common denominator. Then add the whole numbers.

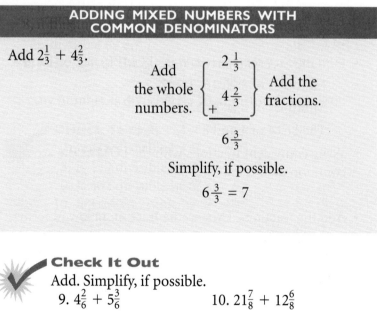

ADDING MIXED NUMBERS WITH COMMON DENOMINATORS

Add $2\frac{1}{3} + 4\frac{2}{3}$.

Add the whole numbers. $\left.\begin{array}{r} 2\frac{1}{3} \\ 4\frac{2}{3} \\ + \end{array}\right\}$ Add the fractions.

$6\frac{3}{3}$

Simplify, if possible.

$6\frac{3}{3} = 7$

Check It Out

Add. Simplify, if possible.

9. $4\frac{2}{6} + 5\frac{3}{6}$ 10. $21\frac{7}{8} + 12\frac{6}{8}$

11. $23\frac{7}{10} + 37\frac{3}{10}$

Adding Mixed Numbers with Unlike Denominators

You can use fraction pieces to model the addition of mixed numbers with unlike denominators.

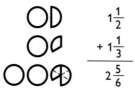

$$1\frac{1}{2}$$
$$+\ 1\frac{1}{3}$$
$$\overline{\quad 2\frac{5}{6}}$$

To add mixed numbers with unlike denominators, you need to write equivalent fractions with a common denominator.

ADDING MIXED NUMBERS WITH UNLIKE DENOMINATORS

Add $2\frac{2}{5} + 3\frac{1}{10}$.

• Write equivalent fractions with a common denominator.

$$2\frac{2}{5} = 2\frac{4}{10} \text{ and } 3\frac{1}{10} = 3\frac{1}{10}$$

• Add.

$$\text{Add the whole numbers.} \left\{ \begin{array}{c} 2\frac{4}{10} \\ + 3\frac{1}{10} \end{array} \right\} \text{Add the fractions.}$$

$$\overline{\quad 5\frac{5}{10}}$$

Simplify, if possible.

$$5\frac{5}{10} = 5\frac{1}{2}$$

Check It Out

Add. Simplify, if possible.

12. $1\frac{5}{8} + 4\frac{3}{4}$

13. $4\frac{5}{12} + 55\frac{3}{4}$

14. $46\frac{1}{2} + 12\frac{7}{8}$

Subtracting Mixed Numbers with Common or Unlike Denominators

You can model the subtraction of *mixed numbers* (p. 105) with unlike denominators.

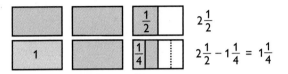

$2\frac{1}{2}$

$2\frac{1}{2} - 1\frac{1}{4} = 1\frac{1}{4}$

To subtract mixed numbers, you need to have or make common denominators.

SUBTRACTING MIXED NUMBERS

Subtract $12\frac{2}{3} - 5\frac{3}{4}$.

- If the denominators are unlike, write equivalent fractions with a common denominator.

$$12\frac{2}{3} = 12\frac{8}{12} \qquad 5\frac{3}{4} = 5\frac{9}{12}$$

- Then subtract.

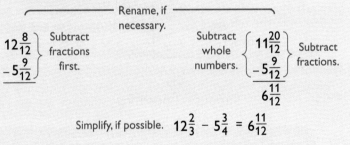

Rename, if necessary.

$12\frac{8}{12}$ } Subtract fractions first.
$-5\frac{9}{12}$

Subtract whole numbers.

$11\frac{20}{12}$ } Subtract fractions.
$-5\frac{9}{12}$
$\overline{6\frac{11}{12}}$

Simplify, if possible. $12\frac{2}{3} - 5\frac{3}{4} = 6\frac{11}{12}$

Check It Out

Subtract. Express your answers in lowest terms.

15. $6\frac{7}{8} - \frac{1}{2}$

16. $32\frac{1}{2} - 16\frac{5}{15}$

17. $30\frac{4}{5} - 12\frac{5}{6}$

18. $26\frac{2}{5} - 17\frac{7}{10}$

Estimating Fraction Sums and Differences

To **estimate** fraction sums and differences, you can use the estimation techniques of rounding or substituting compatible numbers.

ESTIMATING FRACTION SUMS AND DIFFERENCES

Estimate the sum of $7\frac{3}{8} + 8\frac{1}{9} + 5\frac{6}{7}$.

SUBSTITUTE COMPATIBLE NUMBERS

Change each number to a whole number or a mixed number containing $\frac{1}{2}$.

$$7\frac{3}{8} + 8\frac{1}{9} + 5\frac{6}{7}$$
$$\downarrow \qquad \downarrow \qquad \downarrow$$
$$7\frac{1}{2} + 8 \quad + 6 = 21\frac{1}{2}$$

ROUND THE FRACTION PARTS

Round down if the fraction part is less than $\frac{1}{2}$. Round up if the fraction part is greater than or equal to $\frac{1}{2}$.

$$7\frac{3}{8} + 8\frac{1}{9} + 5\frac{6}{7}$$
$$\downarrow \qquad \downarrow \qquad \downarrow$$
$$7 \quad + 8 \quad + 6 = 21$$

Check It Out

Estimate each sum or difference. Use both the compatible numbers method and the rounding method for each problem.

19. $6\frac{1}{4} - 2\frac{5}{6}$
20. $12\frac{1}{8} - 4\frac{3}{4}$

21. $2 + 1\frac{1}{2} + 5\frac{3}{8}$
22. $3\frac{1}{8} + \frac{3}{4} + 4\frac{1}{6}$

2•3 ADDITION AND SUBTRACTION

The Ups and Downs of Stocks

A corporation raises money by selling stocks—certificates that represent shares of ownership in the corporation. The stock page of a newspaper lists the high, low, and ending prices of the stock for the previous day. It also shows the overall fractional amount by which the price changed. A plus (+) sign indicates that the value of the stock increased; a minus (−) sign indicates the value decreased.

Suppose you see a listing on the stock page that shows the closing price of a stock was $21\frac{3}{4}$ with $+\frac{1}{4}$ next to it. What do those fractions mean? First, it tells you that the price of the stock was $21\frac{3}{4}$ dollars or $21.75. The $+\frac{1}{4}$ means the price went up $\frac{1}{4}$ of a dollar from the day before. Since $\frac{1}{4} \times \$1.00 = \$0.25$, the stock went up 25¢. To find the percent increase in the price of the stock, you have to first determine the original price of the stock. The stock went up $\frac{1}{4}$, so the original price is $21\frac{3}{4} - \frac{1}{4} = 21\frac{1}{2}$. What is the percent of increase to the nearest whole percent? See Hot Solutions for answer.

2·3 EXERCISES

Add or subtract. Express in lowest terms.

1. $\frac{3}{16} + \frac{5}{16}$

2. $\frac{5}{25} + \frac{6}{25}$

3. $\frac{3}{8} + \frac{7}{8}$

4. $\frac{15}{29} - \frac{14}{29}$

5. $\frac{25}{60} - \frac{15}{60}$

Add or subtract. Express in lowest terms.

6. $\frac{3}{5} + \frac{6}{9}$

7. $\frac{4}{12} + \frac{2}{15}$

8. $\frac{5}{18} + \frac{1}{6}$

9. $\frac{7}{9} - \frac{2}{3}$

10. $\frac{5}{9} - \frac{1}{3}$

Estimate each sum or difference.

11. $7\frac{9}{10} + 8\frac{3}{4}$

12. $4\frac{5}{6} - 3\frac{3}{4}$

13. $2\frac{9}{10} + 8\frac{3}{5} + 1\frac{1}{2}$

14. $13\frac{4}{5} - 6\frac{1}{9}$

15. $5\frac{1}{3} + 2\frac{7}{8} + 6\frac{1}{4}$

Add or subtract. Simplify, if possible.

16. $7\frac{4}{10} - 3\frac{1}{10}$

17. $3\frac{3}{8} - 1\frac{2}{8}$

18. $13\frac{4}{5} + 12\frac{2}{5}$

19. $24\frac{6}{11} + 11\frac{5}{11}$

20. $22\frac{2}{7} + 11\frac{4}{7}$

Add. Simplify, if possible.

21. $3\frac{1}{2} + 5\frac{1}{4}$

22. $17\frac{1}{3} + 23\frac{1}{6}$

23. $26\frac{3}{4} + 5\frac{1}{2}$

24. $21\frac{7}{10} + 16\frac{3}{5}$

Subtract. Simplify, if possible.

25. $6\frac{1}{5} - 1\frac{9}{10}$

26. $19\frac{1}{4} - 1\frac{1}{2}$

27. $48\frac{1}{3} - 19\frac{11}{12}$

28. $55\frac{3}{8} - 26\frac{2}{7}$

29. Maria is painting her bedroom. She has $4\frac{2}{3}$ gal of paint. She needs $\frac{3}{4}$ gal for the trim and $3\frac{1}{2}$ gal for the walls. Does she have enough paint to paint her bedroom?

30. Maria has a piece of molding $22\frac{5}{8}$ ft long. She used $8\frac{2}{3}$ ft for one wall in the bedroom. Does she have enough molding for the other wall which is $12\frac{1}{2}$ ft long?

2·4 Multiplication and Division of Fractions

Multiplying Fractions

You know that 2×2 means "2 groups of 2." Multiplying fractions involves the same concept: $2 \times \frac{1}{2}$ means "2 groups of $\frac{1}{2}$." You will find it helpful to think of *times* as meaning *of*.

One group of $\frac{1}{2}$

$1 \times \frac{1}{2} = \frac{1}{2}$

Two groups of $\frac{1}{2}$

$2 \times \frac{1}{2} = \frac{2}{2} = 1$

Three groups of $\frac{1}{2}$

$3 \times \frac{1}{2} = \frac{3}{2} = 1\frac{1}{2}$

The same is true when you are multiplying a fraction by a fraction. For example, $\frac{1}{4} \times \frac{1}{2}$ means you would actually be finding $\frac{1}{4}$ of $\frac{1}{2}$.

$\frac{1}{4} \times \frac{1}{2} = \frac{1}{8}$

When you are not using models to multiply fractions, you multiply the numerators and then the denominators. There is no need to find a common denominator.

$$\frac{1}{3} \times \frac{1}{4} = \frac{1}{12}$$

MULTIPLYING FRACTIONS

Multiply $\frac{4}{5}$ and $\frac{5}{6}$.

$\frac{4}{5} \times \frac{5}{6}$ • Write mixed numbers, if any, as improper fractions (p. 106).

$\frac{4}{5} \times \frac{5}{6} = \frac{20}{30}$ ← • Multiply the numerators.
           ← • Multiply the denominators.

$\frac{20 \div 10}{30 \div 10} = \frac{2}{3}$ • Write the product in lowest terms, if necessary.

$\frac{4}{5} \times \frac{5}{6} = \frac{2}{3}$

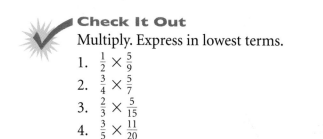

Check It Out

Multiply. Express in lowest terms.

1. $\frac{1}{2} \times \frac{5}{9}$

2. $\frac{3}{4} \times \frac{5}{7}$

3. $\frac{2}{3} \times \frac{5}{15}$

4. $\frac{3}{5} \times \frac{11}{20}$

Shortcut for Multiplying Fractions

You can use a shortcut when you multiply fractions. Instead of multiplying across and then writing the product in lowest terms, you can cancel **factors** first.

CANCELING FACTORS

Multiply $\frac{2}{5}$ and $\frac{10}{14}$.

$\frac{2}{5} \times \frac{10}{14}$ • Write mixed numbers, if any, as improper fractions.

$= \frac{2}{\underset{1}{5}} \times \frac{\overset{1}{2} \cdot \overset{1}{5}}{\underset{1}{2} \cdot 7}$ • Cancel factors, if you can.

$= \frac{2}{1} \times \frac{1}{7} = \frac{2}{7}$ • Multiply.

$= \frac{2}{7}$ • Write the product in lowest terms, if necessary.

Check It Out

Multiply, canceling factors. Express in lowest terms.

5. $\frac{4}{9} \times \frac{3}{8}$

6. $\frac{3}{5} \times \frac{15}{21}$

7. $\frac{12}{15} \times \frac{4}{9}$

8. $\frac{7}{8} \times \frac{16}{21}$

Finding the Reciprocal of a Number

To find the **reciprocal** of a number, you switch the numerator and the denominator.

Number	Reciprocal
$\frac{4}{5}$	$\frac{5}{4}$
$3 = \frac{3}{1}$	$\frac{1}{3}$
$6\frac{1}{2} = \frac{13}{2}$	$\frac{2}{13}$

When you multiply a number by its reciprocal, the product is 1.

$$\frac{2}{5} \times \frac{5}{2} = \frac{10}{10} = 1$$

The number 0 does not have a reciprocal.

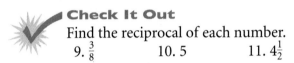

Check It Out

Find the reciprocal of each number.

9. $\frac{3}{8}$ 10. 5   11. $4\frac{1}{2}$

Multiplying Mixed Numbers

You can use what you know about multiplying fractions to help you multiply mixed numbers. To multiply mixed numbers, you rewrite them as improper fractions.

MULTIPLYING MIXED NUMBERS

Multiply $2\frac{1}{3} \times 1\frac{1}{4}$.

• Write the mixed numbers as improper fractions.

$$2\frac{1}{3} \times 1\frac{1}{4} = \frac{7}{3} \times \frac{5}{4}$$

• Cancel factors, if you can, and then multiply the fractions.

$$\frac{7}{3} \times \frac{5}{4} = \frac{35}{12}$$

• Change to a mixed number and reduce to *lowest terms* (p. 102), if necessary.

$$\frac{35}{12} = 2\frac{11}{12}$$

Check It Out
Multiply. Reduce to lowest terms.

12. $2\frac{2}{5} \times 3\frac{2}{6}$ 13. $4\frac{5}{9} \times 2\frac{1}{16}$

14. $15\frac{2}{3} \times 4\frac{5}{8}$ 15. $16\frac{1}{2} \times 4\frac{3}{4}$

Dividing Fractions

When you divide a fraction by a fraction, such as $\frac{1}{3} \div \frac{1}{6}$, you are really finding out how many $\frac{1}{6}$'s are in $\frac{1}{3}$.

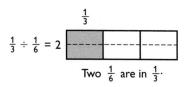

$$\frac{1}{3} \div \frac{1}{6} = 2$$

Two $\frac{1}{6}$ are in $\frac{1}{3}$.

To divide fractions, you replace the divisor with its reciprocal and then multiply to get your answer.

$$\frac{1}{3} \div \frac{1}{6} = \frac{1}{3} \times \frac{6}{1} = 2$$

DIVIDING FRACTIONS

Divide $\frac{3}{4} \div \frac{9}{10}$.

• Replace the divisor with its reciprocal and multiply.

$$\frac{3}{4} \div \frac{9}{10} = \frac{3}{4} \times \frac{10}{9}$$

• Cancel factors.

$$\frac{\overset{1}{\cancel{3}}}{2 \times \cancel{2}_{1}} \times \frac{\overset{1}{\cancel{2}} \times 5}{\cancel{3} \times 3}_{1} = \frac{1}{2} \times \frac{5}{3}$$

• Multiply the fractions.

$$\frac{1}{2} \times \frac{5}{3} = \frac{5}{6}$$

 Check It Out
Divide. Reduce to lowest terms.

16. $\frac{3}{4} \div \frac{3}{5}$ 17. $\frac{5}{7} \div \frac{1}{2}$ 18. $\frac{7}{9} \div \frac{1}{8}$

Dividing Mixed Numbers

When you divide $3\frac{3}{4}$ by $1\frac{1}{4}$ you are actually finding out how many sets of $1\frac{1}{4}$'s are in $3\frac{3}{4}$.

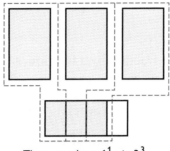

There are three $1\frac{1}{4}$s in $3\frac{3}{4}$.

Dividing mixed numbers is similar to dividing fractions. Before you begin the process, you need to change the *mixed numbers* to *improper fractions* (p. 106).

$$3\frac{3}{4} \div 1\frac{1}{4} = \frac{15}{4} \div \frac{5}{4} = \frac{15}{4} \times \frac{4}{5} = \frac{60}{20} = 3$$

DIVIDING MIXED NUMBERS

Solve $2\frac{1}{2} \div 1\frac{1}{3}$.

- Write the mixed numbers as improper fractions.

  $2\frac{1}{2} \div 1\frac{1}{3} = \frac{5}{2} \div \frac{4}{3}$

- Replace the divisor with its reciprocal.

  The reciprocal of $\frac{4}{3}$ is $\frac{3}{4}$.

- Multiply the fractions. Reduce to lowest terms, if necessary.

  $\frac{5}{2} \times \frac{3}{4} = \frac{15}{8} = 1\frac{7}{8}$

Check It Out

Divide. Express fractions in lowest terms.

19. $1\frac{1}{8} \div \frac{3}{4}$

20. $\frac{16}{2} \div 1\frac{1}{2}$

21. $4\frac{3}{4} \div 6\frac{1}{3}$

EXERCISES

Multiply. Reduce to lowest terms.

1. $\frac{2}{5} \times \frac{7}{9}$ 2. $\frac{3}{8} \times \frac{3}{5}$

3. $\frac{6}{7} \times \frac{7}{8}$ 4. $\frac{3}{4} \times \frac{5}{11}$

5. $\frac{4}{7} \times \frac{21}{24}$

Find the reciprocal.

6. $\frac{5}{7}$ 7. $5\frac{1}{2}$

8. 4 9. $6\frac{2}{3}$

10. $7\frac{5}{6}$ 11. 27

Multiply. Reduce to lowest terms.

12. $5\frac{1}{8} \times 12\frac{2}{7}$ 13. $3\frac{3}{4} \times 16\frac{4}{5}$

14. $11\frac{1}{2} \times 4\frac{1}{6}$ 15. $10\frac{2}{9} \times 2\frac{13}{16}$

16. $6\frac{5}{12} \times 4\frac{4}{9}$ 17. $8\frac{4}{5} \times 5\frac{5}{8}$

Divide. Reduce to lowest terms.

18. $\frac{1}{5} \div \frac{2}{3}$ 19. $\frac{4}{9} \div \frac{11}{15}$

20. $\frac{3}{8} \div \frac{12}{21}$ 21. $\frac{13}{19} \div \frac{26}{27}$

22. $\frac{21}{26} \div \frac{12}{13}$

Divide. Reduce to lowest terms.

23. $5\frac{5}{6} \div 7\frac{7}{9}$ 24. $3\frac{3}{5} \div 2\frac{2}{17}$

25. $12\frac{2}{7} \div 2\frac{13}{15}$ 26. $7\frac{1}{3} \div 6\frac{1}{9}$

27. $4\frac{4}{5} \div 3\frac{4}{5}$ 28. $3\frac{1}{3} \div 1\frac{2}{3}$

29. Each dinner at the Shady Tree Truck Stop is served with $\frac{1}{2}$ cup corn. If there are about 4 cups of corn in a pound, about how many dinners could be served using 16 pounds of corn?

30. Jamie is making chocolate cupcakes. The recipe calls for $\frac{2}{3}$ cup cocoa. She wants to make $\frac{1}{2}$ of the recipe. How much cocoa will she need?

2·5 Naming and Ordering Decimals

Decimal Place Value: Tenths and Hundredths

You can use what you know about **place value** for whole numbers when you read and write decimals.

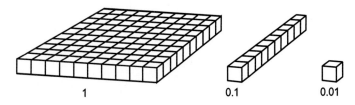

1   0.1   0.01

The base-ten blocks show that:
One whole is ten times greater than one tenth (0.1).
One tenth (0.1) is ten times greater than one hundredth (0.01).

You can use a place-value chart to help you read and write decimal numbers.

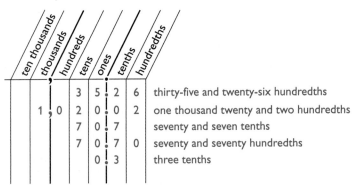

ten thousands	thousands	hundreds	tens	ones	tenths	hundredths	
		3	5	2	6		thirty-five and twenty-six hundredths
	1,0	2	0	0	2		one thousand twenty and two hundredths
		7	0	7			seventy and seven tenths
		7	0	7	0		seventy and seventy hundredths
			0	3			three tenths

You can read the decimal by reading the whole number to the left of the decimal point as usual. You say "and" for the decimal point. Then find the place of the last decimal digit and use it to name the decimal.

You can write a decimal by writing the whole number, putting a decimal point, and then placing the last digit of the decimal number in the place that names it.

1,000 + 20 + 0.02 is 1,020.02 written in expanded notation. The place-value chart can help you write decimals in expanded notation. You write each nonzero place as a number and add them together.

Check It Out
Write the decimal.
1. nine tenths
2. fifty-five hundredths
3. seven and eighteen hundredths
4. five and three hundredths

Decimal Place Value: Thousandths

Thousandths is used as an accurate measurement in sports statistics and scientific studies. The number 1 is 1,000 times one thousandth, and one hundredth is equal to ten thousandths ($\frac{1}{100} = \frac{10}{1,000}$).

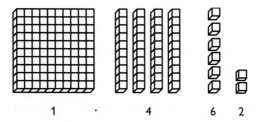

The base-ten blocks model the decimal number 1.462.

- Read the number:
    "one *and* four hundred sixty-two thousandths"
- The decimal number in expanded form is:
    1 + 0.4 + 0.06 + 0.002

Check It Out
Write the decimal in expanded form.
5. 0.634
6. 3.221
7. 0.077

Naming Decimals Greater Than and Less Than One

Decimal numbers are based on units of ten.

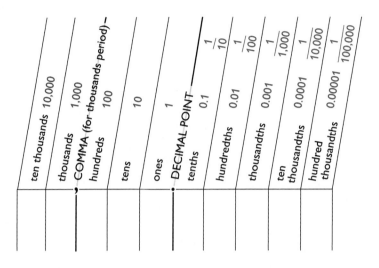

The chart shows the value for some of the digits of a decimal. You can use a place-value chart to help you name decimals greater than and less than one.

NAMING DECIMALS GREATER THAN AND LESS THAN ONE

Find the value of the digits in the decimal number 45.6317.

• Values to the left of the decimal point are greater than one.

45 means 4 tens and 5 ones.

• Read the decimal. The word name of the decimal is determined by the place value of the digit in last place.

The last digit (7) is in the ten-thousandths place.

45.6317 is read as forty-five *and* six thousand three hundred seventeen ten thousandths.

Check It Out
Use the place-value chart to tell what each blue digit means. Then write the numbers in words.
8. 5.633
9. 0.045
10. 6.0074
11. 0.00271

Comparing Decimals

Zeros can be added to the right of the decimal in the following manner without changing the value of the number.

1.045 = 1.0450 = 1.04500 = 1.04500...

To compare decimals, you can compare their place value.

COMPARING DECIMALS

Compare 18.4053 and 18.4063.

- Start at the left. Find the first place where the numbers are different.

    18.4053 and 18.4063

    The thousandths place is different.

- Compare the digits that are different.

    5 < 6

- The numbers compare the same way the digits compare.

18.4053 < 18.4063

Check It Out
Write <, >, or = for each ☐.
12. 37.5 ☐ 37.60   13. 15.336 ☐ 15.636
14. 0.0018 ☐ 0.0015

Ordering Decimals

To write decimals from least to greatest and vice versa, you need to first compare the numbers two at a time.

Order the decimals: 1.123; 0.123; 1.13.

- Compare the numbers two at a time.

    1.123 > 0.123
    1.13 > 1.123

- List the decimals from least to greatest.

    0.123; 1.123; 1.13

Check It Out
Write in order from least to greatest.
15. 4.0146; 40.146; 4.1406
16. 8.073; 8.373; 8; 83.037
17. 0.522; 0.552; 0.52112; 0.5512

Rounding Decimals

Rounding decimals is similar to rounding whole numbers.
Round 13.046 to the nearest hundredth.

- Find the rounding place.   13.046
                                    ↑
                              hundredths

- Look at the digit to the right of the rounding place.   13.046

- If it is less than 5, leave the digit in the rounding place unchanged. If it is greater than or equal to 5, increase the digit in the rounding place by 1.   6 > 5

- Write the rounded number.
    13.05

13.046 rounded to the nearest hundredth is 13.05.

Check It Out
Round each decimal to the nearest hundredth.
18. 1.656   19. 226.948
20. 7.399   21. 8.594

 **2·5 EXERCISES**

Write the decimal.
1. four and twenty-six hundredths
2. five tenths
3. seven hundred fifty-six ten thousandths

Write the decimal in expanded form.
4. seventy-six thousandths
5. seventy-five and one hundred thirty-four thousandths

Give the value of each blue digit.
6. 34.2**4**1
7. 4.34**6**1
8. 0.129**6**
9. 24.1**4**

Compare. Use $<$, $>$, or $=$ for each ☐.
10. 14.0990 ☐ 14.11
11. 13.46400 ☐ 13.46
12. 8.1394 ☐ 8.2
13. 0.664 ☐ 0.674

List in order from least to greatest.
14. 0.707; 0.070; 0.70; 0.777
15. 5.722; 5.272; 5.277; 5.217
16. 4.75; 0.75; 0.775; 77.5

Round each decimal to the indicated place.
17. 1.7432 tenths
18. 49.096 hundredths

19. Five girls are entered into a gymnastic competition in which the highest possible score is 10.0. On the floor routine, Rita scored 9.3, Minh 9.4, Sujey 9.9, and Sonja 9.8. What score does Aisha have to receive in order to win the competition?

20. Based on the chart below, which bank offers the best rate of interest for savings accounts?

Savings Banks	Interest
First Federal	7.25
Western Trust	7.125
National Savings	7.15
South Central	7.1

2·6 Decimal Operations

Adding and Subtracting Decimals

Adding and subtracting decimals is similar to adding and subtracting whole numbers

ADDING AND SUBTRACTING DECIMALS	
Add 3.65 + 0.5 + 22.45.	
• Line up the decimal points.	3.65 0.5 +22.45
• Add or subtract the place farthest right. Rename, if necessary.	$\overset{1}{3.65}$ 0.5 +22.45 0
• Add or subtract the next place left. Rename, if necessary.	$\overset{1\ 1}{3.65}$ 0.5 +22.45 60
• Continue through the whole numbers. Place the decimal point in the result.	3.65 0.5 +22.45 26.60

Check It Out

Solve.
1. 18.68 + 47.30 + 22.9
2. 16.8 + 5.99 + 39.126
3. 6.77 − 0.64
4. 47.026 − 0.743

Estimating Decimal Sums and Differences

One way that you can estimate decimal sums and differences is to use compatible numbers. Compatible numbers are numbers close to the real numbers in the problem, but easier to work with mentally.

ESTIMATING DECIMAL SUMS AND DIFFERENCES

Estimate the sum of 1.344 + 8.744.

* Replace the numbers with compatible numbers.

  1.344 → 1

  8.744 → 9

* Add the numbers.

  1 + 9 = 10

  1.344 + 8.744 is about 10.

Estimate the difference of 18.572 − 7.231.

* Replace the numbers with compatible numbers.

  18.572 → 18

  7.231 → 7

* Subtract the compatible numbers.

  18 − 7 = 11

  18.572 − 7.231 is about 11.

Check It Out

Estimate each sum or difference.

5. 7.64 + 4.33

6. 12.4 − 8.3

7. 19.144 − 4.66

8. 2.66 + 3.14 + 6.54

Multiplying Decimals

Multiplying decimals is much the same as multiplying whole numbers. You can model the multiplication of decimals with a 10 by 10 grid. Each tiny square is equal to one hundredth.

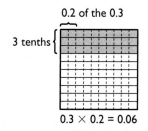

0.2 of the 0.3

3 tenths

$0.3 \times 0.2 = 0.06$

MULTIPLYING DECIMALS

Multiply 24.5×0.07.

- Multiply as with whole numbers.

$$\begin{array}{r} 24.5 \\ \times 0.07 \\ \hline \end{array} \qquad \begin{array}{r} 245 \\ \times\ 7 \\ \hline 1715 \end{array}$$

- Add the number of decimal places for the factors.

$$\begin{array}{r} 24.5 \rightarrow \text{1 decimal place} \\ \times 0.07 \rightarrow \text{2 decimal places} \\ \hline 1715 \quad 1 + 2 = \text{3 decimal places in the answer} \end{array}$$

- Place the decimal point in the product.

$$\begin{array}{r} 24.5 \rightarrow \text{1 decimal place} \\ \times 0.07 \rightarrow \text{2 decimal places} \\ \hline 1.715 \rightarrow \text{3 decimal places} \end{array}$$

$24.5 \times 0.07 = 1.715$

 **Check It Out**

9. 2.8×1.68 10. 33.566×3.4

(sidebar) **2·6 DECIMAL OPERATIONS**

Multiplying Decimals with Zeros in the Product

Sometimes when you are multiplying decimals you need to add zeros in the product.

ZEROS IN THE PRODUCT

Multiply 0.9×0.0456.

• Multiply as with whole numbers. Count decimal places in factors to find the places you need in the product.

$$\begin{array}{r} 0.0456 \\ \times\ 0.9 \\ \hline \end{array} \qquad \begin{array}{r} 456 \\ \times\ 9 \\ \hline 4104 \end{array} \qquad \text{You need 5 decimal places.}$$

• Add zeros in the product, as necessary.

Because 5 decimal places are needed in the product, write one zero to the left of the 4.

$0.0456 \times 0.9 = 0.04104$

 Check It Out

11. 0.051×0.033 12. 0.881×0.055

Estimating Decimal Products

To estimate decimal products, you can replace given numbers with compatible numbers. Compatible numbers are estimates you choose because they are easier to work with mentally.

Estimate the product of 37.3×48.5.

• Replace the factors with compatible numbers.

$37.3 \rightarrow 40$
$48.5 \rightarrow 50$

• Multiply mentally.

$40 \times 50 = 2,000$

Check It Out

Estimate each product, using compatible numbers.

13. 34.84 × 6.6

14. 43.87 × 10.63

Olympic Decimals

In Olympic gymnastics, the competitors perform a set of specific events. Scoring is based on a ten-point scale, where ten is a perfect score. Marks may be given in decimal numbers. After the high and low scores have been eliminated, the remaining marks are averaged.

For some of the events, gymnasts are judged on their technical merit and for composition and style. Marks for technical merit are based on the difficulty and variety of the routine and the skills of the gymnasts. Marks for composition and style are based on the originality and artistry of the routine.

	Technical Merit	Composition and Style
USA	9.4	9.8
China	9.6	9.7
France	9.3	9.9
Germany	9.5	9.6
Australia	9.6	9.7
Canada	9.5	9.6
Japan	9.7	9.8
Russia	9.6	9.5
Sweden	9.4	9.7
England	9.6	9.7

Use these marks to determine the mean Olympic scores for technical merit and for composition and style. (*Hint:* To find an average, add the scores and divide by the number of scores.) See Hot Solutions for answers.

Dividing Decimals

Dividing decimals is similar to dividing whole numbers. You can use a model to help you understand how to divide decimals. For example, $0.8 \div 0.2$ means, how many groups of 0.2 in 0.8? There are 4 groups of 0.2 in 0.8, so $0.8 \div 0.2 = 4$.

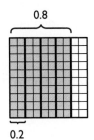

0.8

0.2

DIVIDING DECIMALS

Divide $0.592 \div 1.6$.

- Multiply the divisor by a power of ten, so it is a whole number.

$$1.6 \times 10 = 16$$

- Multiply the dividend by the same power of ten.

$$0.592 \times 10 = 5.92$$

- Place the decimal point in the quotient.

$$16.\overline{)5.92}^{\,\cdot}$$

- Divide.

$$
\begin{array}{r}
0.37 \\
16.\overline{)5.92} \\
4\,8 \\
\hline
112 \\
112 \\
\hline
\end{array}
$$

$$0.592 \div 1.6 = 0.37$$

Check It Out
Divide.
15. $10.5 \div 2.1$ 16. $0.0936 \div 0.02$
17. $3.024 \div 0.06$ 18. $3.68 \div 0.08$

Rounding Decimal Quotients

You can use a calculator to divide decimals. Then you can follow these steps to round the quotient.

Divide $8.3 \div 3.6$. Round to the nearest hundredth.

- Use your calculator to divide.
  8.3 $\boxed{\div}$ 3.6 $\boxed{=}$ 2.3055555

- To round the quotient, look at one place to the right of the rounding place.
  2.305

- If the digit to the right of the rounding place is 5 or above, round up. If the digit to the right of the rounding place is less than 5, the digit to be rounded stays the same.
  $5 = 5$, so 2.305555556 rounded to the nearest hundredth is 2.31.

Check It Out
Use a calculator to find each quotient. Round to the nearest hundredth.
19. $0.509 \div 0.7$
20. $0.1438 \div 0.56$
21. $0.2817 \div 0.47$

2·6 EXERCISES

Estimate each sum or difference.
1. 4.64 + 2.44
2. 7.09 − 4.7
3. 6.666 + 0.34
4. 4.976 + 3.224
5. 12.86 − 7.0064

Add.
6. 224.2 + 3.82
7. 55.12 + 11.65
8. 10.84 + 174.99
9. 8.0217 + 0.71
10. 1.9 + 6 + 2.5433

Subtract.
11. 24 − 10.698
12. 32.034 − 0.649
13. 487.1 − 3.64
14. 53.44 − 17.844
15. 11.66 − 4.0032

Multiply.
16. 0.5 × 5.533
17. 11.5 × 23.33
18. 0.13 × 0.03
19. 39.12 × 0.5494
20. 0.47 × 0.81

Divide.
21. 273.5 ÷ 20.25
22. 29.3 ÷ 0.4
23. 76.5 ÷ 25.5
24. 38.13 ÷ 8.2

Use a calculator to divide. Round the quotient to the nearest hundredth.
25. 583.5 ÷ 13.2
26. 798.46 ÷ 92.3
27. 56.22 ÷ 0.28
28. 0.226 ÷ 0.365

29. The school's record in the field day relay race was 45.78 sec. This year the record was broken by 0.19 sec. What was the new record time this year?

30. Arthur delivers pizzas for $4.75 an hour. Last week he worked 43 hr. How much did he earn?

2·6 EXERCISES

2·7 Meaning of Percent

Naming Percents

Percent is a **ratio** that compares a number with 100. Percent means *per hundred* and is represented by the symbol %.

You can use graph paper to model percents. There are 100 squares in a 10 by 10 grid of graph paper. So the grid can be used to represent 100%. Since percent means how many out of 100, it is easy to tell what percent of the 100-square grid is shaded.

25 of 100 are blue   10 of 100 are red
(25% blue).   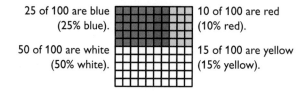   (10% red).

50 of 100 are white   15 of 100 are yellow
(50% white).   (15% yellow).

Check It Out

Give the percents for the number of squares that are shaded and the number of squares that are not shaded.

1.   2.   3.

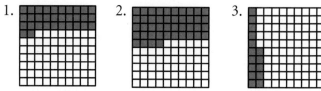

Understanding the Meaning of Percent

Any *ratio* with 100 as the second number can be expressed in three ways. You can write the ratio as a fraction, a decimal, and a percent.

A quarter is 25% of $1.00. You can express a quarter as 25¢, $0.25, $\frac{1}{4}$ of a dollar, $\frac{25}{100}$, and 25%.

One way to think about percents is to become very comfortable with a few. You build what you know about percents based on these few **benchmarks.** You can use these benchmarks to help you estimate percents of other things.

None				Half		All

$0 \; \frac{1}{100} \quad \frac{1}{10} \qquad \frac{1}{4} \qquad\qquad \frac{1}{2} \qquad\qquad \frac{3}{4} \qquad\qquad$ 100%

0.01 0.10   0.25   0.50   0.75

1% 10%   25%   50%   75%

ESTIMATING PERCENTS

Estimate 47% of 60.

- Choose a benchmark, or combination of benchmarks, close to the target percent.

    47% is close to 50%.

- Find the fraction or decimal equivalent to the benchmark percent.

    $50\% = \frac{1}{2}$

- Use the benchmark equivalent to estimate the percent.

    $\frac{1}{2}$ of 60 is 30.

47% of 60 is about 30.

Check It Out

Use fractional benchmarks to estimate the percents.

4. 34% of 70   5. 45% of 80
6. 67% of 95   7. 85% of 32

Using Mental Math to Estimate Percents

You can use fractional or decimal benchmarks in real-life situations to help you quickly estimate the percent of something, such as a tip in a restaurant.

USING MENTAL MATH TO ESTIMATE PERCENT

Estimate a 20% tip for a bill of $15.40.

- Round to a convenient number.

    $15.40 rounds to $15.00

- Think of the percent as a benchmark.

    20% = 0.20

- Multiply mentally.

    $0.20 \times 15.00 = (0.10 \times 15.00) \times 2 = (1.5) \times 2 = \$3.00$

The tip is about $3.00.

Check It Out

Estimate the amount of each tip.

8. 10% of $14.55   9. 23% of $16

10. 47% of $110

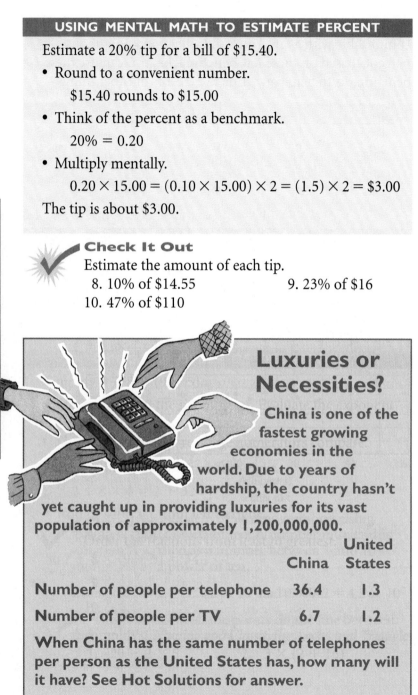

Luxuries or Necessities?

China is one of the fastest growing economies in the world. Due to years of hardship, the country hasn't yet caught up in providing luxuries for its vast population of approximately 1,200,000,000.

	China	United States
Number of people per telephone	36.4	1.3
Number of people per TV	6.7	1.2

When China has the same number of telephones per person as the United States has, how many will it have? See Hot Solutions for answer.

 EXERCISES

Write the percent for the amount that is shaded and for the amount that is not shaded.

1.   2. 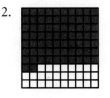   3.

Write each ratio as a fraction, decimal, and percent.
4. 8 to 100
5. 23 to 100
6. 59 to 100

Use fractional benchmarks to estimate the percent of each number.
7. 27% of 60
8. 49% of 300
9. 11% of 75
10. 74% of 80

Use mental math to estimate each percent.
11. 15% of $45
12. 20% of $29
13. 10% of $79
14. 25% of $69
15. 6% of $35

2·7 EXERCISES

2·8 Using and Finding Percents

Finding a Percent of a Number

There are several ways that you can find the percent of a number. You can use decimals or fractions. To find the percent of a number, you must first change the percent to a decimal or a fraction. Sometimes it is easier to change to a decimal representation and other times to a fractional one.

To find 50% of 80, you can use either the fraction method or the decimal method.

FINDING THE PERCENT OF A NUMBER: TWO METHODS

Find 50% of 80.

DECIMAL METHOD

- Change the percent to an equivalent decimal.

    $50\% = 0.5$

- Multiply.

    $0.5 \times 80 = 40$

FRACTION METHOD

- Change the percent to a fraction in lowest terms.

    $50\% = \frac{50}{100} = \frac{1}{2}$

- Multiply.

    $\frac{1}{2} \times 80 = 40$

50% of 80 is equal to 40.

Check It Out

Give the percent of each number.
1. 55% of 35
2. 94% of 600
3. 22% of 55
4. 71% of 36

Finding the Percent of a Number:
Proportion Method

You can use **proportions** to help you find the percent of a number.

USE A PROPORTION TO FIND A PERCENT OF A NUMBER

Forest works in a skateboard store. He receives a commission (part of the sales) of 15% on his sales. Last month he sold $1,400 worth of skateboards, helmets, knee pads, and elbow pads. What was his commission?

- Use a proportion to find the percent of a number.

P = Part (of the base or total) R = Rate (percentage)

B = Base (total) $\frac{P}{R} = \frac{B}{100}$

- Identify the given items before trying to find the unknown.

P is	R is	B is.
unknown call n.	15%.	$1400.

- Set up the proportion.

$$\frac{P}{R} = \frac{B}{100} \qquad \frac{n}{15} = \frac{1400}{100}$$

- Cross multiply.

$$100 \times n = 1400 \times 15 \rightarrow 100n = 21{,}000$$

- Divide both sides of the equation by the coefficient of n.

$$100n = 21{,}000 \qquad\qquad n = \$210$$

Forest received a commission of $210.

2·8 USING AND FINDING PERCENTS

> ### Check It Out
>
> Use a proportion to find the percent of each number.
>
> 5. 56% of 65   6. 12% of 93
> 7. 67% of 139   8. 49% of 400

Finding Percent and Base

Finding what percent a number is of another number and finding what number is a certain percent of another number can be confusing. Setting up and solving a proportion can make the process less confusing.

Use the ratio $\frac{P}{B} = \frac{R}{100}$, where P = Part (of base), B = Base (total), and R = Rate (percentage).

FINDING THE PERCENT

What percent of 30 is 6?

• Set up a proportion, using this form:

$$\frac{\text{Part}}{\text{Base}} = \frac{\text{Percent}}{100}$$

$$\frac{6}{30} = \frac{n}{100}$$

(The number after the word *of* is the base.)

• Show the cross products of the proportion.

$$100 \times 6 = 30 \times n$$

• Find the products.

$$600 = 30n$$

• Divide both sides of the equation by the coefficient of n.

$$\frac{600}{30} = \frac{30n}{30}$$

$$n = 20$$

6 is 20% of 30.

 Check It Out

Solve.

9. What percent of 240 is 60?

10. What percent of 500 is 75?

11. What percent of 60 is 3?

12. What percent of 44 is 66?

FINDING THE BASE

60 is 48% of what number?

- Set up a percent proportion using this form:

$$\frac{\text{Part}}{\text{Base}} = \frac{\text{Percent}}{100}$$

$$\frac{60}{n} = \frac{48}{100}$$

(The phrase *what number* after the word *of* is the base.)

- Show the cross products of the proportion.

$$60 \times 100 = 48 \times n$$

- Find the products.

$$6000 = 48n$$

- Divide both sides of the equation by the coefficient of n.

$$\frac{6000}{48} = \frac{48n}{48}$$

$$n = 125$$

60 is 48% of 125.

Check It Out

Solve. Round the quotient to the nearest hundredth.
13. 54 is 50% of what number?
14. 16 is 80% of what number?
15. 35 is 150% of what number?
16. 74 is 8% of what number?

Percent of Increase or Decrease

Sometimes it is helpful to keep a record of your monthly expenses. This record allows you to see the actual percent of increase or decrease in your spending. You can make a chart to record expenses.

2·8 USING AND FINDING PERCENTS

Expenses	January	February	Amount of Increase or Decrease	Percent of Increase or Decrease
Lunches	$47	$35	$12	
School Supplies	$15	$7	$8	53%
Snacks	$20	$33	$13	65%
Movies	$8	$12		
Miscellaneous	$14	$10	$4	29%
Total	$104	$97		

You can use a calculator to find the percent of increase or decrease.

FINDING THE PERCENT OF INCREASE

During January $8 was spent on movies. In February the amount spent on movies was $12.

- On a calculator, key in the following:

  new amount $\boxed{-}$ original amount $\boxed{=}$ amount of increase

  $12 \boxed{-} 8 \boxed{=} \boxed{\qquad 4.}$

- Leave the amount of increase on the display.

  $\boxed{\qquad 4.}$

- Use your calculator to divide the amount of increase by the original amount.

  amount of increase $\boxed{\div}$ original amount $\boxed{=}$ percent of increase

  $\boxed{\qquad 4.} \boxed{\div} 8 \boxed{=} \boxed{\qquad 0.5}$

- Round and convert to a percent.

  $0.5 = 50\%$

The percent of increase from $8 to $12 is 50%.

Check It Out
Use a calculator to find the percent of increase.
17. 15 to 29   18. 23 to 64
19. 6 to 88   20. 5 to 25

FINDING THE PERCENT OF DECREASE

During January $104 total was spent. In February $97 total was spent. You can use a calculator to find the percent of decrease in amounts spent in January and February.

- On a calculator, key in the following:

| original amount | − | new amount | = | amount of decrease |

104 − 97 = [7.]

- Leave the amount of decrease on display.

[7.]

- Use your calculator to divide the amount of decrease by the original amount.

| amount of decrease | ÷ | original amount | = | percent of decrease |

[7.] ÷ 104 = [0.0673076]

- Round the quotient to the nearest hundredth and convert from decimal to percent.

0.0673076 = 0.07 = 7%

The percent of decrease is 7%.

Check It Out

Use a calculator to find the percent of decrease. Round the quotient to the nearest whole percent.

21. 65 to 21   22. 42 to 18
23. 156 to 122   24. 143 to 60

Discounts and Sale Prices

A **discount** is the amount that an item's price is reduced from the regular price. The sale price is the regular price minus the discount. Discount stores have regular prices below the suggested retail price. To find discounts and resulting sale prices, you can use percents.

2·8 USING AND FINDING PERCENTS

This television has a regular price of $199.99. It is on sale for 20% off the regular price. How much money will you save by buying the item on sale?

regularly $199.⁹⁹

You can use a calculator to help you find the discount and the resulting sale price of an item.

2•8 USING AND FINDING PERCENTS

FINDING DISCOUNTS AND SALE PRICES

The regular price of an item is $199.99. The percent off is 20%. Find the discount and the sale price.

- Use a calculator to multiply the regular price times the discount percent. This will give you the amount of discount.

  regular price ☒ discount percent ＝ discount
  199.99 ☒ 20 ％ ＝ [**39.998**]

- If necessary, round the discount to the nearest hundredth.

  $39.998 → $40.00   The discount is $40.00.

- Use a calculator to subtract the discount from the regular price. This will give you the sale price.

  regular price ⊟ discount ＝ sale price
  199.99 ⊟ 40.00 ＝ [**159.99**]

The sale price is $159.99.

Check It Out

Use a calculator to find the discount and sale price.
25. Regular price: $75, Discount percent: 25%
26. Regular price: $180, Discount percent: 15%

Estimating a Percent of a Number

You can use what you know about compatible numbers and simple fractions to estimate a percent of a number. You can use the table to help you estimate the percent of a number.

Percent	1%	5%	10%	20%	25%	33⅓%	50%	66⅔%	75%	100%
Fraction	$\frac{1}{100}$	$\frac{1}{20}$	$\frac{1}{10}$	$\frac{1}{5}$	$\frac{1}{4}$	$\frac{1}{3}$	$\frac{1}{2}$	$\frac{2}{3}$	$\frac{3}{4}$	1

ESTIMATING A PERCENT OF A NUMBER

Estimate 17% of 46.

- Find the percent that is closest to the percent you are asked to find.

  17% is about 20%.

- Find the fractional equivalent for the percent.

  20% is equivalent to $\frac{1}{5}$.

- Find a compatible number for the number you are asked to find the percent of.

  46 is about 50.

- Use the fraction to find the percent.

  $\frac{1}{5}$ of 50 is 10.

17% of 46 is about 10.

Check It Out

Use compatible numbers to estimate.

27. 67% of 150   28. 35% of 6
29. 27% of 54   30. 32% of 89

Finding Simple Interest

When you have a savings account, the bank pays you for the use of your money. When you take out a loan, you pay the bank for the use of their money. In both situations, the payment is called the *interest*. The amount of money you borrow or save is called the *principal*.

You want to borrow $2,000 at 5% interest for 2 years. To find out how much interest you would owe, you can use the formula $I = P \times R \times T$. The chart below can help you to understand the formula.

P	Principal—the amount of money you borrow or save
R	Interest Rate—a percent of the principal you pay or earn
T	Time—the length of time you borrow or save
I	Total Interest—interest you pay or earn for the entire time
A	Amount—total amount (principal plus interest) you pay or earn

FINDING SIMPLE INTEREST

You can use a calculator to help you find the interest that you would owe on $2,000 borrowed at 5% interest for 2 yr.

- Multiply the principal (P) by the interest rate (R) by the time (T) to find the interest you will pay (I).

    $P \times R \times T = I$

    2000 $\boxed{\times}$ 5 $\boxed{\%}$ $\boxed{\times}$ 2 $\boxed{=}$ $\boxed{200.}$

    $200 is the interest.

- Add the principal and the interest to find the total amount (A) you will pay.

    $P + I = A$

    2000 $\boxed{+}$ 200 $\boxed{=}$ $\boxed{2200.}$

$2,200 is the total amount of money to be paid back to the lender.

Check It Out

Find the interest (I) and the total amount (A).

31. $P = \$650$
    $R = 11\%$
    $T = 3$ years

32. $P = \$2,400$
    $R = 14\%$
    $T = 2\frac{1}{2}$ years

2·8 EXERCISES

Find the percent of each number.
1. 7% of 34
2. 34% of 135
3. 85% of 73
4. 3% of 12.4

Solve.
5. What percent of 500 is 35?
6. What percent of 84 is 147?
7. 52 is what percent of 78?
8. What percent of 126 is 42?

Solve. Round to the nearest hundredth.
9. 38% of what number is 28?
10. 23% of what number is 13?
11. 97% of what number is 22?
12. 65% of what number is 34.2?

Find the percent of increase or decrease to the nearest percent.
13. 7 to 9
14. 56 to 22
15. 21 to 16
16. 13 to 21

Find the discount and the sale price.
17. Regular price: $79
    Discount: 15%
18. Regular price: $229
    Discount: 25%

19. Regular price: $189
    Discount: 30%
20. Regular price: $359
    Discount: 45%

Find the interest (I) and the final amount (A). Use a calculator.
21. $P = \$7{,}500$
    $R = 5.5\%$
    $T = 1$ year
22. $P = \$1{,}100$
    $R = 6\%$
    $T = 2$ years

Estimate the percent of each number.
23. 12% of 72
24. 29% of 185
25. 79% of 65

2·9 Fraction, Decimal, and Percent Relationships

Percents and Fractions

Percents and fractions both describe a ratio out of 100. The chart below will help you to understand the relationship between percents and fractions.

Percent	Fraction
50 out of 100 = 50%	$\frac{50}{100} = \frac{1}{2}$
$33\frac{1}{3}$ out of 100 = $33\frac{1}{3}$%	$\frac{33.\overline{3}}{100} = \frac{1}{3}$
25 out of 100 = 25%	$\frac{25}{100} = \frac{1}{4}$
20 out of 100 = 20%	$\frac{20}{100} = \frac{1}{5}$
10 out of 100 = 10%	$\frac{10}{100} = \frac{1}{10}$
1 out of 100 = 1%	$\frac{1}{100}$
$66\frac{2}{3}$ out of 100 = $66\frac{2}{3}$%	$\frac{66.\overline{6}}{100} = \frac{2}{3}$
75 out of 100 = 75%	$\frac{75}{100} = \frac{3}{4}$

You can write fractions as percents and percents as fractions.

CONVERTING A FRACTION TO A PERCENT

Use a proportion to express $\frac{2}{5}$ as a percent.

- Set up a proportion. $\frac{2}{5} = \frac{n}{100}$
- Solve the proportion. $5n = 2 \times 100$

    $n = 40$

- Express as a percent. $\frac{2}{5} = 40\%$

2·9 RELATIONSHIPS

Check It Out

Change each fraction to a percent. Round to the nearest whole percent.

1. $\frac{11}{20}$ 2. $\frac{4}{10}$
3. $\frac{6}{8}$ 4. $\frac{3}{7}$

Changing Percents to Fractions

To change from a percent to a fraction, write the percent as the numerator of a fraction with a denominator of 100, and express in lowest terms.

CHANGING PERCENTS TO FRACTIONS

Express 45% as a fraction.

- Change the percent directly to a fraction with a denominator of 100. The number of the percent becomes the numerator of the fraction.

  $45\% = \frac{45}{100}$

- Express the fraction in *lowest terms* (p. 102).

  $\frac{45}{100} = \frac{9}{20}$

45% expressed as a fraction in lowest terms is $\frac{9}{20}$.

2•9 RELATIONSHIPS

Check It Out

Convert each percent to a fraction in lowest terms.

5. 16%   6. 4%
7. 38%   8. 72%

Changing Mixed Number Percents to Fractions

To change the mixed-number percent $15\frac{1}{4}\%$ to a fraction, first change the mixed number to an *improper fraction* (p. 106).

- Change the mixed number to an improper fraction.

  $15\frac{1}{4}\% = \frac{61}{4}\%$

- Multiply the percent by $\frac{1}{100}$.

  $\frac{61}{4} \times \frac{1}{100} = \frac{61}{400}$

- Simplify, if possible.

  $15\frac{1}{4}\% = \frac{61}{400}$

Check It Out

Convert each mixed number percent to a fraction expressed in lowest terms.

9. $24\frac{1}{2}\%$ 10. $16\frac{3}{4}\%$

11. $121\frac{1}{8}\%$

Honesty Pays

David Hacker, a cabdriver, found a wallet in the back seat of the cab that contained $25,000— about a year's salary for him.

The owner's name was in the wallet and Hacker remembered where he had dropped him off. He went straight to the hotel and found the man. The owner, a businessman, had already realized he had lost his wallet and figured he would never see it again. He didn't believe anyone would be that honest! On the spot, he handed the cabdriver fifty $100 bills.

What percent of the money did Hacker receive as a reward? See Hot Solutions for answer.

Percents and Decimals

Percents can be expressed as decimals and decimals can be expressed as percents. *Percent* means part of a hundred or hundredths.

CHANGING DECIMALS TO PERCENTS

Change 0.9 to a percent.
- Multiply the decimal by 100.

    $0.9 \times 100 = 90$

- Add the percent sign to the product.

    $0.9 = 90\%$

A Shortcut for Changing Decimals to Percents
Change 0.9 to a percent.

- Move the decimal point two places to the right. Add zeros, if necessary.

    $0.9 \longrightarrow 0.90.$

- Add the percent sign.

    90%

    So 0.9 = 90%.

Check It Out
Write each decimal as a percent.
12. 0.45
13. 0.606
14. 0.019
15. 2.5

Changing Percents to Decimals

Because *percent* means part of a hundred, percents can be converted directly to decimals.

CHANGING PERCENTS TO DECIMALS

Change 6% to a decimal.

- Express the percent as a fraction with 100 as the denominator.

    $6\% = \frac{6}{100}$

- Change the fraction to a decimal by dividing the numerator by the denominator.

    $6 \div 100 = 0.06$

    $6\% = 0.06$

A Shortcut for Changing Percents to Decimals

Change 6% to a decimal.

- Move the decimal point two places to the left.

    $6\% \rightarrow .\underset{\smile}{\,6.}$

- Add zeros, if necessary.

    $6\% = 0.06$

Check It Out

Express each percent as a decimal.

16. 54%
17. 190%
18. 4%
19. 29%

Fractions and Decimals

Fractions can be written as either **terminating** or **repeating decimals.**

Fractions	Decimals	Terminating or Repeating
$\frac{1}{2}$	0.5	terminating
$\frac{1}{3}$	0.3333333...	repeating
$\frac{1}{6}$	0.166666...	repeating
$\frac{2}{3}$	0.666666...	repeating
$\frac{3}{5}$	0.6	terminating

CHANGING FRACTIONS TO DECIMALS

Write $\frac{2}{5}$ as a decimal.

- Divide the numerator of the fraction by the denominator.

  $2 \div 5 = 0.4$

The remainder is zero. 0.4 is a terminating decimal.

Write $\frac{2}{3}$ as a decimal.

- Divide the numerator of the fraction by the denominator.

  $2 \div 3 = 0.666666...$

  The decimal is a repeating decimal.

- Place a bar over the digit that repeats.

  $0.6\overline{6}$ or $0.\overline{6}$

$\frac{2}{3} = 0.6\overline{6}$. It is a repeating decimal.

Check It Out

Use a calculator to find a decimal for each fraction.

20. $\frac{4}{5}$ 21. $\frac{5}{16}$ 22. $\frac{5}{9}$

2•9 RELATIONSHIPS

CHANGING DECIMALS TO FRACTIONS

Write the decimal 0.24 as a fraction.

- Write the decimal as a fraction.

$$0.24 = \frac{24}{100}$$

- Express the fraction in *lowest terms* (p. 102).

$$\frac{24}{100} = \frac{24 \div 4}{100 \div 4} = \frac{6}{25}$$

So $0.24 = \frac{6}{25}$.

 Check It Out

Write each decimal as a fraction.

23. 0.225
24. 0.5375
25. 0.36

**EXERCISES**

Change each fraction to a percent.

1. $\frac{1}{5}$ 2. $\frac{3}{25}$

3. $\frac{1}{100}$ 4. $\frac{13}{50}$

Change each percent to a fraction in lowest terms.

5. 28%   6. 64%

7. 125%   8. 87%

Write each decimal as a percent.

9. 0.9   10. 0.27

11. 0.114   12. 0.55

13. 3.7

Write each percent as a decimal.

14. 38%   15. 13.6%   16. 19%

17. 5%   18. 43.2%

Change each fraction to a decimal. Use a bar to show repeating digits.

19. $\frac{1}{5}$ 20. $\frac{2}{9}$ 21. $\frac{3}{16}$

22. $\frac{4}{9}$ 23. $\frac{9}{10}$

Write each decimal as a fraction in lowest terms.

24. 0.05   25. 0.005   26. 10.3

27. 0.875   28. 0.6

29. Bargain Barn is offering CD players at 50% off the regular price of $149.95. Larry's Lowest is offering CD players at $\frac{1}{3}$ off the regular price of $119.95. Which store has the better buy?

30. One survey at Franklin Middle School said 24% of the sixth grade students named basketball as their favorite sport. Another survey said $\frac{6}{25}$ of the sixth grade students named basketball as their favorite sport. Could both surveys be correct? Explain.

2.9 EXERCISES

What have you learned? You can use the problems and list of words below to see what you have learned in this chapter. To find out more about a particular problem or word, refer to the boldfaced topic number (for example, **1•2**).

Problem Set

1. Miguel bought 0.8 kg of grapes at $0.55 a kilogram and 15 grapefruits at $0.69 each. How much did he spend? **2•6**

2. Hadas bought $12\frac{3}{4}$ yd of curtain material. She is planning to make curtains for two windows. One window requires $6\frac{1}{2}$ yd of material and the other window requires $5\frac{3}{4}$ yd of material. Does Hadas have enough material to make both curtains? **2•3**

3. Jacob increased his word-processing speed from 40 to 48 words per minute. What is his percent of increase in speed? **2•8**

4. Which fraction is equivalent to $\frac{16}{24}$? **2•1**
   A. $\frac{2}{3}$ B. $\frac{8}{20}$ C. $\frac{4}{5}$ D. $\frac{6}{4}$

5. Which fraction is greater, $\frac{1}{16}$ or $\frac{2}{19}$? **2•2**

Add or subtract. Write your answers in lowest terms. **2•3**

6. $\frac{3}{5} + \frac{5}{9}$ 7. $4\frac{1}{7} - 2\frac{3}{4}$

8. $6 - 1\frac{2}{3}$ 9. $7\frac{1}{9} + 2\frac{7}{8}$

10. Write the improper fraction $\frac{16}{5}$ as a mixed number. **2•1**

Multiply or divide. Write your answers in lowest terms. **2•4**

11. $\frac{4}{7} \times \frac{6}{7}$ 12. $\frac{2}{3} \div 7\frac{1}{4}$

13. $4\frac{1}{4} \times \frac{3}{4}$ 14. $5\frac{1}{4} \div 2\frac{1}{5}$

15. Give the place value of 5 in 432.159. **2•5**

16. Write in expanded form: 4.613. **2•5**

17. Write as a decimal: three hundred and sixty-six thousandths. **2•5**

18. Write the following numbers in order from least to greatest: 0.660; 0.060; 0.066; 0.606. **2•5**

Find each answer as indicated. **2•6**

19. $12.344 + 2.89$

20. $14.66 - 0.487$

21. 34.89×0.0076

22. $0.86 \div 0.22$

Use a calculator to answer items 23–25. Round to the nearest tenth. **2•8**

23. What is 53% of 244?

24. Find 154% of 50.

25. 17 is what percent of 20?

Write each decimal as a percent. **2•9**

26. 0.65

27. 0.05

Write each fraction as a percent. **2•9**

28. $\frac{3}{8}$

29. $\frac{7}{20}$

Write each percent as a decimal. **2•9**

30. 16%

31. 3%

Write each percent as a fraction in lowest terms. **2•9**

32. 36%

33. 248%

WRITE DEFINITIONS FOR THE FOLLOWING WORDS.

hot **words**

benchmark **2•7**
common denominator **2•2**
cross product **2•1**
denominator **2•1**
discount **2•8**
equivalent **2•1**

equivalent fractions **2•1**
estimate **2•3**
factor **2•4**
fraction **2•1**
greatest common factor **2•1**
improper fraction **2•1**
least common multiple **2•2**
mixed number **2•1**

numerator **2•1**
percent **2•7**
place value **2•5**
product **2•4**
proportion **2•8**
ratio **2•7**
repeating decimal **2•9**
terminating decimal **2•9**
whole number **2•1**

Powers and Roots

$2^3 = 8$

What do you already know?

You can use the problems and list of words below to see what you already know about this chapter. The answers to the problems are in Hot Solutions at the back of the book, and the definitions of the words are in Hot Words at the front of the book. You can find out more about a particular problem or word by referring to the boldfaced number (for example, **3•2**).

Problem Set

Write each multiplication using an exponent. **3•1**
1. $7 \times 7 \times 7 \times 7 \times 7$
2. $a \times a \times a \times a \times a \times a \times a \times a$
3. $4 \times 4 \times 4$
4. $x \times x$
5. $3 \times 3 \times 3 \times 3$

Evaluate each square. **3•1**
6. 2^2
7. 5^2
8. 10^2
9. 7^2
10. 12^2

Evaluate each cube. **3•1**
11. 2^3
12. 4^3
13. 10^3
14. 7^3
15. 1^3

Evaluate each power of 10. **3•1**
16. 10^2
17. 10^6
18. 10^{10}
19. 10^7
20. 10^1

Evaluate each square root. **3•2**
21. $\sqrt{9}$
22. $\sqrt{25}$
23. $\sqrt{144}$
24. $\sqrt{64}$
25. $\sqrt{4}$

Estimate each square root between two consecutive numbers. **3•2**
26. $\sqrt{20}$
27. $\sqrt{45}$
28. $\sqrt{5}$
29. $\sqrt{75}$
30. $\sqrt{3}$

Estimate each square root to the nearest thousandth. **3•2**
31. $\sqrt{5}$
32. $\sqrt{20}$
33. $\sqrt{50}$
34. $\sqrt{83}$
35. $\sqrt{53}$

CHAPTER 3		
hot **words**	cube **3•1**	square **3•1**
	exponent **3•1**	square root **3•2**
	factor **3•1**	volume **3•1**
area **3•1**	perfect square **3•2**	
base **3•1**	power **3•1**	

WHAT DO YOU KNOW?

3·1 Powers and Exponents

Exponents

Multiplication is the shortcut for showing a repeated addition: $4 \times 6 = 4 + 4 + 4 + 4 + 4 + 4$. A shortcut for showing the repeated multiplication $4 \times 4 \times 4 \times 4 \times 4 \times 4$ is to write 4^6. The 4 is the factor to be multiplied, called the **base**. The 6 is the **exponent**, which tells how many times the base is to be multiplied. The expression can be read as "4 to the sixth **power**." When you write an exponent, it is written slightly higher than the base and is usually a little smaller.

MULTIPLICATION USING EXPONENTS

Write the multiplication $3 \times 3 \times 3 \times 3$ using an exponent.

• Check that the same **factor** is being used in the multiplication.

   All the factors are 3.

• Count the number of times 3 is being multiplied.

   There are 4 factors of 3.

• Write the multiplication using an exponent.

Since the factor 3 is being multiplied 4 times, write 3^4.

Check It Out

Write each multiplication using an exponent.
1. $8 \times 8 \times 8 \times 8$
2. $3 \times 3 \times 3 \times 3 \times 3 \times 3 \times 3$
3. $x \times x \times x$
4. $y \times y \times y \times y \times y$

Evaluating the Square of a Number

The **square** of a number means to apply the exponent 2 to a base. The square of 3, then, is 3^2. To evaluate 3^2, identify 3 as the base and 2 as the exponent. Remember that the exponent tells you how many times to use the base as a factor. So 3^2 means to use 3 as a factor 2 times:

$$3^2 = 3 \times 3 = 9$$

The expression 3^2 can be read as "3 to the second power." It can also be read as "3 squared."

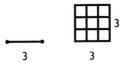

When a square is made from a segment whose length is 3, the **area** of the square is $3 \times 3 = 3^2 = 9$.

EVALUATING THE SQUARE OF A NUMBER

Evaluate 7^2.

- Identify the base and the exponent.

    The base is 7 and the exponent is 2.

- Write the expression as a multiplication.

    $7^2 = 7 \times 7$

- Evaluate.

    $7 \times 7 = 49$

Check It Out

Evaluate each square.

5. 4^2 6. 5^2

7. 8 squared   8. 6 squared

Evaluating the Cube of a Number

The **cube** of a number means to apply the exponent 3 to a base. The cube of 2, then, is 2^3. Evaluating cubes is very similar to evaluating squares. For example, if you wanted to evaluate 2^3, notice that 2 is the base and 3 is the exponent. Remember, the exponent tells you how many times to use the base as a factor. So 2^3 means to use 2 as a factor 3 times:

$$2^3 = 2 \times 2 \times 2 = 8$$

The expression 2^3 can be read as "2 to the third power." It can also be read as "2 cubed."

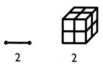

2   2

When a cube has edges of length 2, the **volume** of the cube is $2 \times 2 \times 2 = 2^3 = 8$.

EVALUATING THE CUBE OF A NUMBER

Evaluate 2^3.

- Identify the base and the exponent.

    The base is 2 and the exponent is 3.

- Write the expression as a multiplication.

    $2^3 = 2 \times 2 \times 2$

- Evaluate.

    $2 \times 2 \times 2 = 8$

You can use a calculator to evaluate powers. You simply use the calculator to multiply the right number of times (p. 379). Or you can use special keys (p. 386).

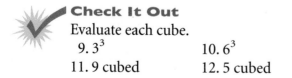

Check It Out

Evaluate each cube.

9. 3^3 10. 6^3

11. 9 cubed   12. 5 cubed

Powers of Ten

Our decimal system is based on 10. For each factor of 10, the decimal point moves one place to the right.

$$3.151 \rightarrow 31.51 \qquad 14.25 \rightarrow 1,425 \qquad 3. \rightarrow 30$$
$$\times 10 \qquad\qquad \times 100 \qquad\qquad \times 10$$

When the decimal point is at the end of a number and the number is multiplied by 10, a zero is added at the end of the number.

Try to discover a pattern for the powers of 10.

Powers	As a Multiplication	Result	Number of Zeros
10^2	10×10	100	2
10^4	$10 \times 10 \times 10 \times 10$	10,000	4
10^5	$10 \times 10 \times 10 \times 10 \times 10$	100,000	5
10^8	$10 \times 10 \times 10 \times 10 \times 10 \times 10 \times 10 \times 10$	100,000,000	8

Notice that the number of zeros after the 1 is the same as the power of 10. This means that, if you want to evaluate 10^7, simply write a 1 followed by 7 zeros: 10,000,000.

 Check It Out
Evaluate each power of 10.
13. 10^3 14. 10^5
15. 10^{10} 16. 10^8

Bugs

Insects are the most successful form of life on Earth. About one million have been classified and named. It is estimated that there are up to four million more. That's not total insects we are talking about; that's different *kinds* of insects!

Insects have adapted to life in every imaginable place. They live in the soil, in the air, on the bodies of plants, animals, and other insects, on the edges of salt lakes, in pools of oil, and in the hot waters of hot springs.

Insects range in length from the stick-insect which can be over 15 inches long to a tiny parasitic fly less than .01 inches in length. Insects show an incredible range of adaptations. The tiny midge can beat its wings 62,000 times a minute. A flea can jump 130 times its height. Ants are so social they have been known to live in colonies with 1,000,000 queens and 300,000,000 workers.

Estimates are that there are 200,000,000 insects for each person on the planet. Given a world population of approximately 6,000,000,000, just how many insects do we share the earth with? Use a calculator to arrive at an estimate. Express the number in scientific notation. See Hot Solutions for answer.

3•1 POWERS AND EXPONENTS

3·1 EXERCISES

Write each multiplication using an exponent.
1. $7 \times 7 \times 7$
2. $6 \times 6 \times 6 \times 6 \times 6 \times 6 \times 6 \times 6$
3. $y \times y \times y \times y \times y \times y$
4. $m \times m \times m \times m \times m \times m \times m \times m \times m \times m$
5. 12×12

Evaluate each square.
6. 2^2
7. 7^2
8. 10^2
9. 1 squared
10. 15 squared

Evaluate each cube.
11. 3^3
12. 8^3
13. 11^3
14. 10 cubed
15. 7 cubed

Evaluate each power of 10.
16. 10^2
17. 10^6
18. 10^{14}

19. What is the area of a square whose sides have a length of 9?
    A. 18
    B. 36
    C. 81
    D. 729
20. What is the volume of a cube whose sides have a length of 5?
    A. 60
    B. 120
    C. 125
    D. 150

Square Roots

In mathematics, certain operations are opposites of each other. That is, one operation "undoes" the other. For example, addition undoes subtraction: $3 - 2 = 1$, so $1 + 2 = 3$. Multiplication undoes division: $6 \div 3 = 2$, so $2 \times 3 = 6$. Finding the **square root** of a number undoes the squaring of that number. You know that 3 squared $= 3^2 = 9$. The square root of 9 is the number that can be multiplied by itself to get 9, which is 3. The symbol for square root is $\sqrt{\ }$. Therefore $\sqrt{9} = 3$.

FINDING THE SQUARE ROOT

Find $\sqrt{25}$.

- Think, what number times itself makes 25?

    $5 \times 5 = 25$

- Find the square root.

    Since $5 \times 5 = 25$, the square root of 25 is 5.

Thus $\sqrt{25} = 5$.

✓ Check It Out

Find each square root.

1. $\sqrt{16}$
2. $\sqrt{25}$
3. $\sqrt{64}$
4. $\sqrt{100}$

Estimating Square Roots

The table shows the first ten **perfect squares** and their square roots.

Perfect square	1	4	9	16	25	36	49	64	81	100
Square root	1	2	3	4	5	6	7	8	9	10

You can estimate the value of a square root by finding the two consecutive numbers that the square root must be between.

ESTIMATING A SQUARE ROOT

Estimate $\sqrt{40}$.

- Identify the perfect squares that 40 is between.
  40 is between 36 and 49.
- Find the square roots of the perfect squares.
  $\sqrt{36} = 6$ and $\sqrt{49} = 7$.
- Estimate the square root.
  $\sqrt{40}$ is between 6 and 7.

Check It Out
Estimate each square root.
5. $\sqrt{20}$ 6. $\sqrt{38}$

Better Estimates of Square Roots

If you want to know a better estimate for the value of a square root, you will want to use a calculator. Most calculators have a key $\boxed{\sqrt{}}$ for finding square roots.

On some calculators, the $\sqrt{}$ function is shown not on a key, but above the $\boxed{x^2}$ key on the calculator's surface. If this is true for your calculator, you should also find a key that has either $\boxed{\text{INV}}$ or $\boxed{\text{2nd}}$ on it. To use the $\sqrt{}$ function, you would press $\boxed{\text{INV}}$ or $\boxed{\text{2nd}}$, then the key with $\sqrt{}$ above it.

ESTIMATING THE SQUARE ROOT OF A NUMBER

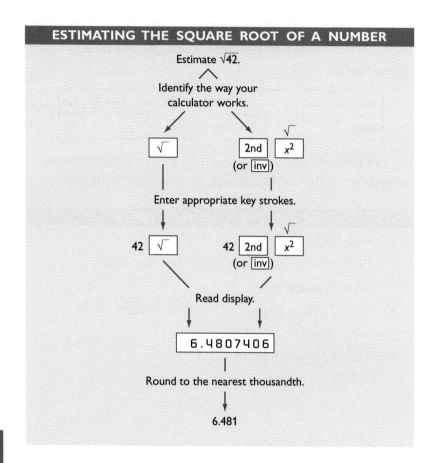

Estimate $\sqrt{42}$.

Identify the way your calculator works.

$\sqrt{}$

$\boxed{\text{2nd}}$ $\boxed{x^2}$ $\sqrt{}$
(or $\boxed{\text{inv}}$)

Enter appropriate key strokes.

42 $\sqrt{}$

42 $\boxed{\text{2nd}}$ $\boxed{x^2}$ $\sqrt{}$
(or $\boxed{\text{inv}}$)

Read display.

$$6.4807406$$

Round to the nearest thousandth.

6.481

Check It Out

Estimate each square root to the nearest thousandth.

7. $\sqrt{2}$ 8. $\sqrt{28}$

EXERCISES

Find each square root.
1. $\sqrt{9}$
2. $\sqrt{49}$
3. $\sqrt{121}$
4. $\sqrt{4}$
5. $\sqrt{144}$

6. $\sqrt{30}$ is between which two numbers?
   A. 3 and 4
   B. 5 and 6
   C. 29 and 31
   D. None of these
7. $\sqrt{72}$ is between which two numbers?
   A. 4 and 5
   B. 8 and 9
   C. 9 and 10
   D. 71 and 73
8. $\sqrt{10}$ is between which two consecutive numbers?
9. $\sqrt{41}$ is between which two consecutive numbers?
10. $\sqrt{105}$ is between which two consecutive numbers?

Estimate each square root to the nearest thousandth.
11. $\sqrt{3}$
12. $\sqrt{15}$
13. $\sqrt{50}$
14. $\sqrt{77}$
15. $\sqrt{108}$

What have you learned?

You can use the problems and list of words below to see what you have learned in this chapter. You can find out more about a particular problem or word by referring to the boldfaced topic number (for example, **3•2**).

Problem Set

Write each multiplication using an exponent. **3•1**

1. $5 \times 5 \times 5 \times 5 \times 5 \times 5 \times 5 \times 5$
2. $m \times m \times m \times m$
3. 9×9
4. $y \times y \times y \times y \times y \times y \times y \times y \times y \times y$
5. $45 \times 45 \times 45 \times 45$

Evaluate each square. **3•1**

6. 3^2
7. 6^2
8. 12^2
9. 8^2
10. 15^2

Evaluate each cube. **3•1**

11. 3^3
12. 6^3
13. 1^3
14. 8^3
15. 2^3

Evaluate each power of 10. **3•1**

16. 10^3
17. 10^5
18. 10^8
19. 10^{13}
20. 10^1

Evaluate each square root. **3•2**

21. $\sqrt{4}$
22. $\sqrt{36}$
23. $\sqrt{121}$
24. $\sqrt{81}$
25. $\sqrt{225}$

Estimate each square root between two consecutive numbers. **3•2**

26. $\sqrt{27}$
27. $\sqrt{8}$
28. $\sqrt{109}$
29. $\sqrt{66}$
30. $\sqrt{5}$

Estimate each square root to the nearest thousandth. **3•2**

31. $\sqrt{11}$
32. $\sqrt{43}$
33. $\sqrt{88}$
34. $\sqrt{6}$
35. $\sqrt{57}$

---

hot **words**

WRITE DEFINITIONS FOR THE FOLLOWING WORDS.

area **3•1**
base **3•1**

cube **3•1**
exponent **3•1**
factor **3•1**
perfect square **3•2**
power **3•1**

square **3•1**
square root **3•2**
volume **3•1**

WHAT HAVE YOU LEARNED?

Data, Statistics, and Probability

What do you already know?

You can use the problems and the list of words that follow to see what you already know about this chapter. The answers to the problems are in Hot Solutions at the back of the book, and the definitions of the words are in Hot Words at the front of the book. To find out more about a particular problem or word, refer to the boldfaced topic number (for example, **4•2**).

Problem Set

1. Jacob surveyed 20 people who were using the pool and asked them if they wanted a new pool. Is this a random sample? **4•1**

2. Sylvia asked 40 people if they planned to vote for the school bond issue. What kind of question did she ask? **4•1**

Use the following graph to answer items 3–5. **4•2**
Vanessa recorded the number of people who used the new school overpass each day.

3. What kind of graph did Vanessa make?

4. On what day did most students use the overpass?

5. Which grades use the overpass the most?

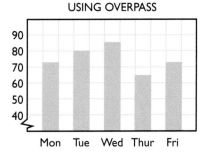

NUMBER OF STUDENTS USING OVERPASS

Use the following data to answer items 6 and 7. The data lists the number of people in line at the bank each time the line was counted.

2 0 4 3 1 2 5 1 0 2 1 5 1 3 6 1

6. Make a frequency graph of the data. **4•2**

7. Make a histogram of the data. **4•2**

8. The weekly wages at the Ice Cream Parlor are $45, $188, $205, $98, and $155. What is the range of wages? **4•4**
9. Find the mean and median of the wages in item 8. **4•4**
10. $P(6, 2) = ?$ **4•5**
11. $C(7, 3) = ?$ **4•5**

12. What is the value of 5!? **4•5**

Use the following information to answer items 13–15. **4•6**
A box contains 40 tennis balls. Eighteen are green and the rest are yellow.
13. One ball is drawn. What is the probability it is yellow? **4•6**
14. Two balls are drawn. What is the probability they are both green? **4•6**
15. A ball is drawn and replaced. Then a second one is drawn. What is the probability that both are yellow? **4•6**

CHAPTER 4

*hot***words**

average **4•4**
box plot **4•2**
circle graph **4•2**
combination **4•5**
dependent events **4•6**
distribution **4•3**
double-bar graph **4•2**
event **4•6**
experimental probability **4•6**
factorial **4•5**
histogram **4•2**

independent events **4•6**
leaf **4•2**
line graph **4•2**
mean **4•4**
median **4•4**
mode **4•4**
normal distribution **4•3**
outcome **4•6**
outcome grid **4•6**
percent **4•2**
permutation **4•5**
population **4•1**
probability **4•6**
probability line **4•6**

random sample **4•1**
range **4•4**
sample **4•1**
sampling with replacement **4•6**
spinner **4•5**
stem **4•2**
stem-and-leaf plot **4•2**
strip graph **4•6**
survey **4•1**
table **4•1**
tally marks **4•1**
theoretical probability **4•6**
tree diagram **4•5**

4·1 Collecting Data

Surveys

Have you ever been asked to name your favorite color? Or asked what kind of music you like? These kinds of questions are often asked in **surveys.** A statistician studies a group of people or objects, called a **population.** They usually get information from a small part of the population, called a **sample.**

In a survey, 150 sixth graders at Kennedy School were chosen at random and asked what kind of pet they had. The following bar graph shows the percent of students who named each type of pet.

In this case, the population is all sixth graders at Kennedy School. The sample is the 150 students who were actually asked to name each type of pet they had.

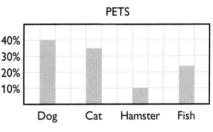

In any survey:
- The population consists of the people or objects about which information is desired.
- The sample consists of the people or objects in the population that are actually studied.

✓ Check It Out
Identify the population and the size of the sample:
1. Sixty students who were signed up for after-school sports were asked if they wanted to have the sports available during the summer.

2. Fifteen wolves on Isle Royale were tagged and let loose.

Random Samples

When you choose a sample to survey for data, be sure the sample is representative of the population. You also should be sure it is a **random sample,** where each person in the population has an equal chance of being included.

Shayna wanted to find out how many of her classmates would like to have a class party at the end of the year. She picked a sample by writing the names of her classmates on cards and drawing 15 cards from a bag. She then asked those 15 classmates whether they wanted to have a class party.

DETERMINING IF A SAMPLE IS RANDOM

Determine if Shayna's sample is random.
- Define the population.

    The population is the students in Shayna's class.
- Define the sample.

    The sample consists of 15 students.
- Determine if the sample is random.

Since every classmate had the same chance of being chosen, the sample is random.

Check It Out

3. How do you think you could select a random sample of your classmates?
4. Suppose you ask 20 people working out at a fitness center which center they prefer. Is this a random sample?

Questionnaires

When you write questions for a survey, it is important to be sure the questions are not biased. That is, the questions should not assume anything or influence the answers. The following two questionnaires are designed to find out what kind of sports you like. The first questionnaire uses biased questions. The second questionnaire uses questions that are not biased.

Survey 1
    A. Do you prefer tame sports like table tennis?
    B. Are you the adventurous type who likes to sky dive?

Survey 2
    A. Do you like to play table tennis?
    B. Do you like to sky dive?

To develop a questionnaire:
* Decide what topic you want to ask about.
* Define a population and decide how to select a sample from that population.
* Develop questions that are not biased.

Check It Out

  5. Why is **A** in Survey 1 biased?

  6. Why is **B** in Survey 2 better than **B** in Survey 1?

  7. Write a question that asks the same thing as the following question but is not biased: Are you a caring person who gives money to charity?

Compiling Data

Once Shayna collected the data from her classmates about a class party, she had to decide how to show the results. As she asked each classmate if they wanted to have a party, she used **tally marks** to tally the answers in a table. The following **table** shows their answers.

Do You Want a Party?	Number of Students
Yes	ℍℍ |||
No	||||
Don't care	|||

To make a table to compile data:
• List the categories or questions in the first column or row.
• Tally the responses in the second column or row.

Check It Out

8. How many students don't care if they have a party?
9. Which answer was given by the greatest number of students?
10. If Shayna uses the survey to decide whether or not to have a party, what should she do? Explain.

4•1 COLLECTING DATA

Chilled to the Bone

Wind carries heat away from the body, increasing the cooling rate. So whenever the wind blows, you feel cooler. If you live in an area where the temperature drops greatly in winter, you know you may feel much, much colder on a blustery winter day than the temperature indicates.

Wind Speed (mi/hr)	Air Temperature (°F)							
	35	30	25	20	15	10	5	0
Calm	35	30	25	20	15	10	5	0
5	32	27	22	16	11	6	0	−5
10	22	16	10	3	−3	−9	−15	−22
15	16	9	2	−5	−11	−18	−25	−31
20	12	4	−3	−10	−17	−24	−31	−39
25	8	1	−7	−15	−22	−29	−36	−44
30	6	−2	−10	−18	−25	−33	−41	−49

This wind-chill table shows the effects of the cooling power of the wind in relation to temperature under calm conditions (no wind). Notice that the wind speed (in miles per hour) is correlated with the air temperature (in degrees Fahrenheit). To determine the wind-chill effect, read across and down to find the entry in the table that matches a given wind speed and temperature.

Listen to or read your local weather report each day for a week or two in the winter. Record the daily average temperature and wind speed. Use the table to determine how chilly it felt each day.

EXERCISES

1. One thousand registered voters were asked what party they prefer. Identify the population and the sample. How big is the sample?
2. Livna wrote the names of 14 classmates on slips of paper and drew five from a bag. Was the sample random?
3. LeRon asked students who ride the bus with him if they participate in clubs at school. Is the sample random?

Are the following questions biased? Explain.
4. How do you get to school?
5. Do you bring a boring lunch from home or buy a school lunch?

Write questions that ask the same thing as the following questions but are not biased.
6. Are you a caring person who recycles?
7. Do you like the uncomfortable chairs in the lunchroom?

Ms. Sandover asked her students which of the following national parks they would like to visit.

National Park	Number of Sixth Graders	Number of Seventh Graders
Yellowstone	卌 II	卌 卌 II
Yosemite	卌 卌 II	IIII
Olympic	卌 III	卌 卌
Grand Canyon	卌 IIII	卌 卌 卌
Glacier	卌 I	卌 II

8. Which park was the most popular? How many students preferred that park?
9. Did more students pick Yellowstone or Olympic?
10. How many students were surveyed?

4·2 Displaying Data

Interpret and Create a Table

You know that statisticians collect data about people or objects. One way to show the data is to use a table. For example, one day during several 15-minute periods, Desrie counted the number of cars that went by her school with the following results.

10 14 13 12 17 18 12 18 18 11 10 13 15 18 17 10 18 10

MAKING A TABLE

Make a table to organize the data about the number of cars.

• Name the first column or row *what* you are counting.

Label the first row *Number of Cars.*

• Tally the amounts for each category in the second column or row.

Number of Cars	10	11	12	13	14	15	16	17	18
Frequency	IIII	I	II	II	I	I		II	ꟷ

• Count the tallies and record the number in the second row.

Number of Cars	10	11	12	13	14	15	16	17	18
Frequency	4	1	2	2	1	1	0	2	5

The most common number of cars was 18. Only once did 11, 14, and 15 cars come by during a 15-minute period.

Check It Out

1. During how many 15-minute periods did Desrie count 10 or more cars?

2. Make a table using the data below of numbers of sponsors who were signed up by students taking part in a walk-a-thon.

4 6 2 5 10 9 8 2 4 6 10 10 4 2 8 9 5 5 5 10 5 9

Interpret a Box Plot

A **box plot** shows data using the middle value of the data and the quartiles, or 25% divisions of the data. The following box plot shows points scored by the Lewistown Larks in winning basketball games.

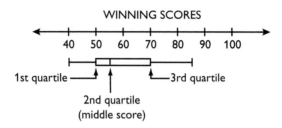

On a box plot, 50% of the scores are above the middle score and 50% are below it. The first-quartile score is the middle score of the bottom half of the scores. The third-quartile score is the middle score in the top half of the scores.

INTERPRETING A BOX PLOT

Interpret the box plot of winning scores (above).

• Identify the high and low values on the plot.

> The high score is at the right: 85. The low score is on the left: 40.

• Find the middle score and the first- and third-quartile scores.

> The middle score is 55, the first-quartile score is 50, and the third-quartile score is 70.

• Give any other information that is available.

> For example, 50% of the scores are between 50 and 70.

4·2 DISPLAYING DATA

Check It Out

Use the following box plot to answer items 3–5.

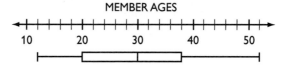

MEMBER AGES

3. What is the age of the oldest person in the club?
4. What percent of the members are younger than 20 years old?
5. What percent of the members are between 30 and 52 years old?

Interpret and Create a Circle Graph

Another way to show data is to use a **circle graph.** Toshi read that in a forest, 25% of the plants is older forest structure, 25% is layered, 25% is understory, 15% is single canopy, and 10% is new growth. He wanted to make a circle graph to show his data.

To make a circle graph:

• Find what **percent** of the whole each part of the data is.

In this case, the percents are given.

• Multiply each percent by 360°, the number of degrees in a circle.

360° × 25% = 90°
360° × 15% = 54°
360° × 10% = 36°

• Draw a circle, measure each central angle (p. 343), and complete the graph.

Toshi made the following graph.

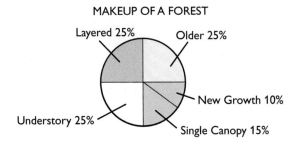

MAKEUP OF A FOREST

Layered 25% Older 25%

New Growth 10%

Understory 25%

Single Canopy 15%

From the graph, you can see that half of the forest is older growth or layered growth.

Check It Out

Use the circle graph to answer items 6 and 7.

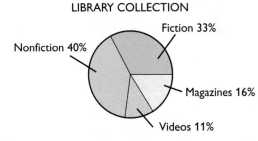

LIBRARY COLLECTION

Fiction 33%

Nonfiction 40%

Magazines 16%

Videos 11%

6. What two categories make up just more than half of the collection?
7. Write a statement about the relationship of the size of the fiction collection to that of the nonfiction.

8. The following pets were entered in a pet show. Make a circle graph to show the entries.
   Dogs: 18   Cats: 20
   Hamsters: 8   Rabbits: 4

Interpret and Create a Frequency Graph

You have used tally marks to show data. Suppose you collect the following information about the number of books your classmates checked out from the library.

5 2 1 0 4 4 2 1 5 2 6 3 2 7 1 5 2 3

You can make a frequency graph by placing X's above a number line.

To make a frequency graph:
• Draw a number line showing the numbers in your data set. In this case, you would draw a number line showing the numbers 0 through 7.
• Place an X to represent each result above the number line for each number you have.
• Title the graph.

In this case, you could call it "Books Checked Out."

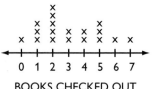

BOOKS CHECKED OUT

You can tell from the frequency graph that students checked out between 0 and 7 library books.

Check It Out

Use the "Books Checked Out" frequency graph to answer items 9 and 10.

9. How many students checked out more than four books?
10. What is the most common number of books checked out?

11. Make a frequency graph to show the number of cars that went by Desrie's school (p. 190).

Interpret a Line Graph

You know that a line graph can be used to show changes in data over time. The following line graph shows the percent of households in the United States with color TV.

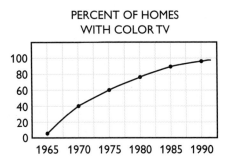

PERCENT OF HOMES
WITH COLOR TV

From the graph, you can see that the percent of households with color TV has steadily increased over the years.

Check It Out

12. In what year did about 60% of households have color TV?

13. Is the following statement true or false? The percent of households with color TV increased more between 1980 and 1990 than between 1965 and 1970.

14. Between what years did the percent of people with color TV double?

Interpret a Stem-and-Leaf Plot

The following numbers show students' scores on a math quiz.

33 27 36 18 30 24 31 33 27 32 27 35 23 40 22 34 28
31 28 28 26 31 28 32 25 29

It is hard to tell much about the scores when they are displayed like this. Another way to show the information is to make a **stem-and-leaf plot.** The following stem-and-leaf plot shows the scores.

```
1 | 8
2 | 2 3 4 5 6 7 7 7 8 8 8 8 9
3 | 0 1 1 1 2 2 3 3 4 5 6
4 | 0
```
    2|2 means 22.

Notice that the tens digits appear in the left-hand column. These are called **stems.** Each digit on the right is called a **leaf.** From looking at the plot, you can tell that most of the students scored from 22 to 36 points.

Check It Out

Use this stem-and-leaf plot showing the ages of people who came into The Big Store in the mall to answer items 15–17.

```
0 | 7 9
1 | 0 2 2 4 5 8
2 | 1 3 4 5 6 7 8 8 8
3 | 0 3 4
```
    3|0 means 30 years old.

15. How many people came into the store?
16. Are more people in their twenties or teens?
17. Three people were the same age. What age is that?

Interpret and Create a Bar Graph

Another type of graph you can use to show data is called a *bar graph*. In this graph, either horizontal or vertical bars are used to show data. Consider the data showing the area of Rhode Island's five counties.

Bristol	25 mi^2	Providence	413 mi^2
Kent	170 mi^2	Washington	333 mi^2
Newport	104 mi^2		

Make a bar graph to show the area of Rhode Island's counties.
- Choose a vertical scale and decide what to place along the horizontal scale.

  In this case, the vertical scale can show square miles in increments of 50 square miles and the horizontal scale can show the county names.
- Above each name draw a bar of the appropriate height.
- Write a title for the graph.

  Title this graph "Area of Rhode Island Counties."
  Your bar graph should look like this:

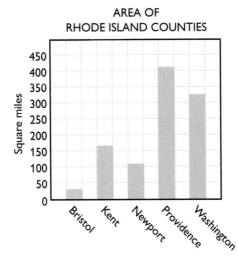

From the graph, you can see that the largest county is Providence County.

4-2 DISPLAYING DATA

Check It Out

Use the bar graph "Area of Rhode Island Counties" (p. 197) to answer items 18 and 19.

18. What is the smallest county?
19. How would the graph be different if every square represented 100 square miles instead of 50?
20. Use the following data to make a bar graph about how students spend their time after school.

Play outdoors: 26　　　Do chores: 8
Talk to friends: 32　　　Watch TV: 18

Interpret a Double-Bar Graph

You know that you can show information in a bar graph. If you want to show information about two or more things, you can use a **double-bar graph.** This graph shows how many sixth and seventh graders liked each school program.

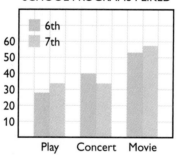

SCHOOL PROGRAMS I LIKED

You can see from the graph that more sixth graders than seventh graders liked the concert.

Check It Out

21. About how many students liked the play the best?
22. Write a conclusion that you can form from this graph.

Graphic Impressions

Humans can live longer than 100 years, but not all animals can live that long. A mouse, for example, has a maximum life span of 3 years, while a toad may live for 36 years.

Both these graphs compare the maximum life span of guppies, giant spiders, and crocodiles.

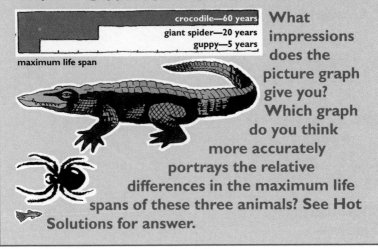

crocodile—60 years
giant spider—20 years
guppy—5 years

maximum life span

What impressions does the picture graph give you? Which graph do you think more accurately portrays the relative differences in the maximum life spans of these three animals? See Hot Solutions for answer.

Interpret and Create a Histogram

A special kind of bar graph that shows frequency of data is called a **histogram.** Several sixth graders told how many grams of fat are in one serving of the cereal they like best. The fat gram results are listed below.

8 9 3 6 4 0 5 6 4 9 4 5 5 0 6 6

MAKING A HISTOGRAM

Create a histogram.

- Make a table showing frequencies.
- Make a bar graph showing the frequencies.
- Title the graph. In this case, you might call it "Grams of Fat in Cereal."

Grams	Tally	Frequency
0	//	2
1		0
2		0
3	/	1
4	///	3
5	///	3
6	////	4
7		0
8	/	1
9	//	2

Your histogram might look like this:

You can see from the histogram that none of the cereals have 1, 2, or 7 grams of fat.

GRAMS OF FAT IN CEREAL

(Histogram: Frequency on vertical axis 0–5, Grams on horizontal axis 0–9)

Check It Out

23. How many sixth graders were surveyed about fat in cereal?
24. Make a histogram from the data showing the number of cars passing Desrie's school (p. 190).

 **EXERCISES**

Use the data about first words in a story to answer items 1–4.

Number of Letters in the First Words in a Story
2 6 3 1 4 4 3 5 2 5 4 5 4 3 1 5 3 2 3 2

1. Make a table and a histogram to show the data about first words in a story.
2. How many words were counted?
3. Make a frequency graph to show the data about first words.
4. Use your frequency graph to describe the number of letters in the words.

5. Kelsey found out that 4 of her friends like art, 6 like math, 5 like science, and 5 like English and social studies. Make a circle graph and write a sentence about it.

6. The following stem-and-leaf plot shows the number of salmon traveling up a fish ladder every hour.

$$
\begin{array}{c|ccccccccc}
7 & 0 & 2 & 4 & 4 & 4 & 4 & 6 & 9 \\
8 & 0 & 1 & 3 & 4 & 4 & 5 & 5 & 6 & 8 & 8 \\
9 & 1 & 1 & 7 \\
\end{array}
$$

8|3 means 83

Draw a conclusion from the plot.

7. The sixth graders recycled 89 pounds of aluminum, the seventh graders recycled 78 pounds, and the eighth graders recycled 92 pounds. Make a bar graph to show this information.

8. The box plot shows the average wait, in seconds, to be helped when calling a toll-free number. What is the shortest wait? 50% of the waits are between 110 and what number of seconds?

TOLL-FREE CALL WAIT TIME

```
   ←——+——+——+——+——+——+——+——+——+——+——→
      50  60  70  80  90  100 110 120 130
      ├——————┌——————————┬——————┐————┤
```

4.3 Analyzing Data

Distribution of Data

Ramon played a game in which he shot a marble to the top of a board and it rolled down the board into one of seven slots. One marble dropped into the first slot, two dropped into the second, three dropped into the third, four dropped into the fourth, three dropped into the fifth, two dropped into the sixth, and one dropped into the seventh. To show the **distribution,** he drew a histogram to show the results of his game.

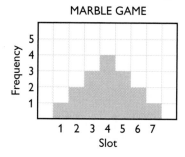

MARBLE GAME

Notice how symmetrical the histogram is. Ramon drew a curve over the histogram by connecting the centers of the bars. The curve he drew illustrates a **normal distribution.**

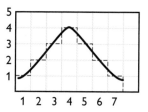

 Check It Out

Tell whether each curve illustrates a normal distribution.

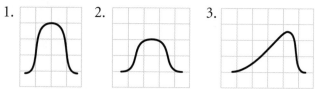

1. 2. 3.

 **EXERCISES**

1. Which of the following is a normal distribution?

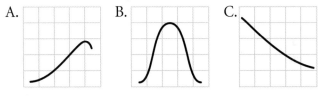

2. Draw a histogram from the following data. Is the distribution normal?

   1 3 5 4 5 2 3 1 3 5 5 4 3 3 3 2 2 4

Birthday Surprise

How likely do you think it is that two people in your class have the same birthday? With 365 days in a year, you might think the chances are very slim. After all, the probability that a person is born on any given day is $\frac{1}{365}$, or about 0.3%.

Try taking a survey. Ask your classmates to write their birthdays on separate slips of paper. Don't forget to write your birthday, too. Collect the slips and see if any birthdays match.

It might surprise you to learn that in a group of 23 people, the chances that two share the same birthday is just a slight bit more than 50%. With 30 people, the likelihood increases to 71%. And with 50 people, you can be 97% sure that two of them were born on the same day.

Statistics

Mei Mei asked 15 friends how much allowance they got. She recorded these amounts:

$1 $1 $2 $3 $3 $3 $3 $5 $6 $6 $7 $7 $8 $8 $42

Mei Mei said her friends typically got $3, but Navid disagreed. He said the typical amount was $5. A third friend, Gabriel, said they were both wrong—the typical amount was $7. Each was correct because each was using a different common measure.

Mean

One measure of data is the **mean.** To find the mean, or **average,** add the allowances together and divide by the total number of allowances being compared.

FINDING THE MEAN

Find the mean allowance.
- Add the amounts.

    $1 + $1 + $2 + $3 + $3 + $3 + $3 + $5 + $6 + $6 + $7 + $7 + $8 + $8 + $42 = $105

- Divide the total by the number of amounts.

    In this case, there are 15 allowances:

    $105 ÷ 15 = $7

The mean allowance is $7. Gabriel used the mean when he said the typical amount was $7.

Check It Out
Find the mean:
1. 12, 15, 63, 12, 24, 34, 23, 15
2. 84, 86, 98, 78, 82, 94
3. 132, 112, 108, 243, 400, 399, 202
4. Ryan earned money baby-sitting. He earned $40, $40, $51, $32, and $22. Find the mean amount he earned.

Median

You see that you can find the mean by adding all the numbers and then dividing by the amount of numbers. Another way to look at numbers is to find the median. The **median** is the middle number in the data when the numbers are arranged in order. Let's look again at the allowances.

$1 $1 $2 $3 $3 $3 $3 $5 $6 $6 $7 $7 $8 $8 $42

FINDING THE MEDIAN

Find the median of the allowances.

- Arrange the data in numerical order from least to greatest or greatest to least.

  Looking at the allowances, we can see they are already arranged in order.

- Find the middle number.

  There are 15 numbers. The middle number is $5 because there are 7 numbers above $5 and 7 numbers below $5.

Navid was using the median when he described the typical allowance.

4•4 STATISTICS

When the number of amounts is even, you can find the median by finding the mean of the two middle numbers. So, to find the median of the numbers 1, 3, 4, 3, 7, and 12, you must find the two numbers in the middle.

FINDING THE MEDIAN OF AN EVEN NUMBER OF DATA

- Arrange the numbers in order from least to greatest or greatest to least.

  1, 3, 3, 4, 7, 12   or   12, 7, 4, 3, 3, 1

- Find the mean of the two middle numbers.

  The two middle numbers are 3 and 4:

  $(3 + 4) \div 2 = 3.5$

The median is 3.5. Half the numbers are greater than 3.5 and half the numbers are less than 3.5.

Check It Out

Find the median:

5. 21, 38, 15, 8, 18, 21, 8

6. 24, 26, 2, 33

7. 90, 96, 68, 184, 176, 86, 116

8. Yeaphana measured the weight of 10 adults, in pounds: 160, 140, 175, 141, 138, 155, 221, 170, 150, and 188. Find the median weight.

4·4 STATISTICS

Mode

You can describe a set of numbers using the mean or by using the median, which is the middle number. Another way to describe a set of numbers is to use the mode. The **mode** is the number in the set that occurs most often. Let's look again at the allowances:

$1, $1, $2, $3, $3, $3, $3, $5, $6, $6, $7, $7, $8, $8, $42

To find the mode, look for the number that appears most frequently.

FINDING THE MODE

Find the mode of the allowances.

- Arrange the numbers in order or make a frequency table of the numbers.

    The numbers are arranged in order above.

- Select the number that appears most frequently.

    The most common allowance is $3.

So Mei Mei was using the mode when she described the typical allowance.

A group of numbers may have no mode or more than one mode. Data that has two modes is called *bimodal.*

 Check It Out

Find the mode:

9. 53, 52, 56, 53, 53, 52, 57, 56
10. 100, 98, 78, 98, 96, 87, 96
11. 12, 14, 14, 16, 21, 15, 14, 13, 20
12. Attendance at the zoo one week was as follows: 34,543; 36,122; 35,032; 36,022; 23,944; 45,023; 50,012.

Range

Another measure used with numbers is the range. The **range** tells how far apart the greatest and least numbers in a set are. Consider the seven tallest buildings in Phoenix, Arizona:

Building Height

Building 1	407 ft
Building 2	483 ft
Building 3	372 ft
Building 4	356 ft
Building 5	361 ft
Building 6	397 ft
Building 7	397 ft

To find the range, you must subtract the shortest height from the greatest.

FINDING THE RANGE

Find the range of the tallest buildings in Phoenix.

• Find the greatest and least values.

   The greatest value is 483 and the least value is 356.

• Subtract.

   $483 - 356 = 127$

The range is 127 feet.

Check It Out

Find the range:

13. 110, 200, 625, 300, 12, 590

14. 24, 35, 76, 99

15. 23°, 6°, 0°, 14°, 25°, 32°

16. For three years the following number of people entered animals in the fair: 228, 612, 558.

Find the mean, median, mode, and range. Round to the nearest tenth.

1. 2, 4, 5, 5, 6, 6, 7, 7, 7, 9
2. 18, 18, 20, 28, 20, 18, 18
3. 14, 13, 14, 15, 16, 17, 23, 14, 16, 20
4. 79, 94, 93, 93, 80, 86, 82, 77, 88, 90, 89, 93

5. Are any of the sets of data in items 1–4 bimodal? Explain.

6. When a cold front went through Lewisville, the temperature dropped from 84° to 38°. What was the range in temperatures?

7. In one week, there were the following number of accidents in Caswell: 1, 1, 3, 2, 5, 2, and 1. Which of the measures (mean, median, or mode) do you think you should use to describe the number of accidents? Explain.

8. Does the median have to be a member of the set of data?

9. The following numbers represent the phone calls received each hour between noon and midnight during one day at a mail-order company.

   13 23 14 12 80 22 14 25 14 17 12 18

   Find the mean, median, and mode of the calls. Which measure best represents the data? Explain.

10. Are you using the mean, median, or mode when you say that half the houses in Sydneyville cost more than $150,000?

4.5 Combinations and Permutations

Tree Diagrams

You often need to be able to count outcomes. For example, suppose you have a **spinner** that is half red and half green. You want to find out all the possible outcomes if you spin the spinner three times. You can make a **tree diagram.**

MAKING A TREE DIAGRAM

How many different results can you get if you spin the red and green spinner three times?

• List what happens on the first spin.

The spinner can show red or green.

First spin results

red

green

• List what happens with the second and third spins.

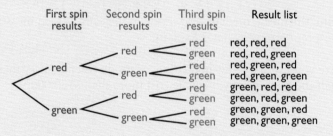

First spin results	Second spin results	Third spin results	Result list
red	red	red	red, red, red
		green	red, red, green
	green	red	red, green, red
		green	red, green, green
green	red	red	green, red, red
		green	green, red, green
	green	red	green, green, red
		green	green, green, green

• Draw lines and list all the possible results.

The results are listed above. There are eight different results.

You can find the number of possibilities by multiplying the number of choices at each step: $2 \times 2 \times 2 = 8$.

Check It Out

Draw a tree diagram for each of the following items. Check by multiplying.

1. Kimiko can choose a sugar cone, a regular cone, or a dish. She can have vanilla, chocolate, or strawberry yogurt. How many possible desserts can she have?
2. Kirti sells round, square, oblong, and flat beads. They come in green, orange, and white. If he separates all the beads by color and shape, how many containers does he need?
3. If you toss three coins, how many possible ways can they land?
4. How many possible ways can Norma climb to Mt. Walker if she goes through Soda Spring?

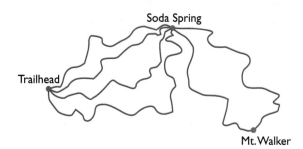

Soda Spring

Trailhead

Mt. Walker

Permutations

The tree diagram shows the different ways in which things can be arranged, or listed. A listing in which the order is important is called a **permutation.** Suppose you want to put three friends, Mariko, Navid, and Taktuk in a line for a picture. How many different arrangements can you find? You can use a tree diagram to show the solution.

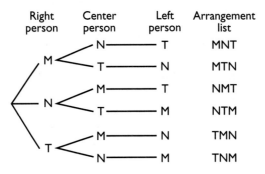

There are three ways to choose the first person and two ways to choose the second. After that, there is only one way to choose the third person. So the total number of permutations is $3 \times 2 \times 1 = 6$. Remember that Mariko, Navid is a different permutation from Navid, Mariko. $P(3, 3)$ represents the number of permutations of three things taken three at a time. Thus $P(3, 3) = 6$.

FINDING PERMUTATIONS

Find $P(7, 4)$.

- Determine how many choices there are for the first, second, third, and fourth places, out of seven people.

  There are 7 choices for the first place, 6 for the second, 5 for the third, and 4 for the fourth.

- Find the product.

  $7 \times 6 \times 5 \times 4 = 840$

So $P(7, 4) = 840$.

Factorial Notation
You saw that to find the number of permutations of six things, you found the product $6 \times 5 \times 4 \times 3 \times 2 \times 1$. The product $6 \times 5 \times 4 \times 3 \times 2 \times 1$ is called 6 **factorial.** The shorthand notation for factorial is an exclamation point. So 6! is called 6 factorial. $6! = 6 \times 5 \times 4 \times 3 \times 2 \times 1 = 720$.

Check It Out
Find each value.
5. $P(5, 3)$
6. $P(6, 2)$
7. At a chess tournament, there are six finalists. In how many ways can six prizes be awarded?
8. One dog out of 12 dogs entered in a show will get first place and one will get second place. In how many different ways can the first- and second-place dogs be chosen?

Find each value. Use a calculator, if available.
9. 7!
10. 5!

Combinations

When you choose two of ten people to play a game of tennis, order is not important. That is, choosing Arturo and Elena to play a game of tennis is the same as choosing Elena and Arturo.

If you want to choose two snacks to pack for a day trip, out of the five choices in your kitchen, the order is not important. Choosing raisins and crackers is the same as choosing crackers and raisins.

Let's look at the two snacks out of five choices.

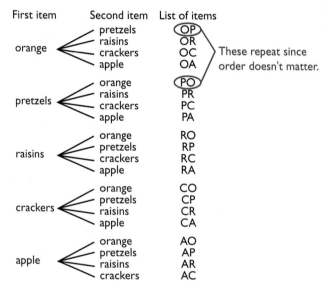

To find the number of **combinations** of five snacks taken two at a time, you start by finding the permutations. You have five ways to choose the first snack and four ways to choose the second, so this is 5 × 4 = 20. But the order doesn't matter, so some combinations were counted too often! You need to divide by the number of different ways two items can be arranged (2!).

$$C(5, 2) = \frac{P(5, 2)}{2!} = \frac{5 \times 4}{2 \times 1} = 10$$

$C(5, 2)$ means the combinations of five items taken two at a time when the order does not matter.

FINDING COMBINATIONS

Find $C(6, 3)$.
- Find $P(6, 3)$.

$P(6, 3) = 6 \times 5 \times 4 = 120$

- Divide by 3!.

$120 \div 3! = 120 \div 6 = 20$

So $C(6, 3) = 20$.

Check It Out
Find each value.

11. $C(7, 2)$
12. $C(10, 6)$
13. How many different combinations of three videos can you choose from eight videos?
14. Are there more combinations or permutations of four dogs from a total of ten? Explain.

Monograms

What are your initials? Do you have anything with your monogram on it? A *monogram* is a design that is made up of one or more letters, usually the initials of a name. Monograms often appear on stationary, towels, shirts, or jewelry.

How many different three-letter monograms can you make with the letters of the alphabet? Use a calculator to compute the total number. Don't forget to allow for repeat letters in the combination. See Hot Solutions for answer.

Find each value.
1. $P(4, 2)$
2. $C(8, 8)$
3. $P(9, 5)$
4. $C(6, 1)$
5. $4! \times 4!$
6. $P(5, 5)$

7. Make a tree diagram to show the results when you spin a spinner containing the numbers 1 through 5 and toss a coin.

8. Eight friends want to play board games in groups of four. How many different ways can they group themselves?

9. Ten people are finalists in a skating competition. In how many ways can the judges award first, second, and third places?

10. Determine if each of the following is a permutation or a combination.
    A. choosing two delegates from 400 students to represent the school at a city conference
    B. electing a president, vice president, and secretary from a club with 18 members

Probability

If you and a friend want to decide who goes first in a game, you might flip a coin. You and your friend have an equal chance of winning the toss. The **probability** of an event is a number from 0 to 1 that measures the chance that an event will occur.

Experimental Probability

The probability of an event is a number between 0 and 1. One way to find the probability of an event is to conduct an experiment. Suppose you want to know the probability of seeing a friend when you ride your bike. You ride your bike 20 times and see a friend 12 times. You can compare the number of times you see a friend to the number of times you ride to find the probability of seeing a friend. In this case, the **experimental probability** that you will see a friend is $\frac{12}{20}$, or $\frac{3}{5}$.

DETERMINING EXPERIMENTAL PROBABILITY

Find the experimental probability of winning a game of chess.

- Conduct an experiment. Record the number of trials and the result of each trial.

   Play chess 10 times and count the wins. Suppose you win 6 times.

- Compare the number of occurrences of one result to the number of trials. That is the probability for that result.

   Compare the number of times you win to the total number of games you play.

The experimental probability of winning a game of chess in this test is $\frac{6}{10}$, or $\frac{3}{5}$.

4·6 PROBABILITY

Check It Out

Slips of paper labeled red, green, yellow, and blue are drawn from a bag containing 40 slips of paper. The results are shown on the circle graph.

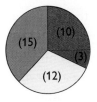

1. Find the experimental probability of getting blue.
2. Find the experimental probability of getting either yellow or red.
3. Toss a coin 50 times. Find the experimental probability of getting a head. Compare your answers with others' answers.

Theoretical Probability

You know that you can find the experimental probability of tossing a head when you toss a coin by doing the experiment and recording the results. You can also find the **theoretical probability** by considering the outcomes of the experiment. The **outcome** of an experiment is a result. The possible outcomes when rolling a standard number cube are the numbers 1–6. An **event** is a specific outcome, such as 5. So the probability of getting a 5 is:

$$\frac{\text{number of ways an event occurs}}{\text{number of outcomes}} = \frac{1 \text{ way to get a } 5}{6 \text{ possible outcomes}} = \frac{1}{6}$$

4·6 PROBABILITY

DETERMINING THEORETICAL PROBABILITY

Find the probability of rolling a two or a three when you roll a number cube containing the numbers 1–6.

- Determine the number of ways the event occurs.

    In this case, the event is getting a two or a three. There are two ways to get a two or three.

- Determine the total number of outcomes. Use a list, multiply, or make a *tree diagram* (p. 210).

    There are six numbers on the cube.

- Use the formula.

$$P(\text{event}) = \frac{\text{number of ways an event occurs}}{\text{number of outcomes}}$$

Find the probability of rolling a two or a three, represented by $P(2 \text{ or } 3)$, by substituting into the formula.

$$P(2 \text{ or } 3) = \frac{2}{6} = \frac{1}{3}$$

The probability of rolling a two or a three is $\frac{1}{3}$.

**Check It Out**

Find each probability. Use the spinner drawing for items 5–6.

4. $P(\text{even number})$ when rolling a 1–6 number cube

5. $P(3)$
6. $P(1, 2, 3, \text{ or } 4)$

7. The letters of the words *United States* are written on slips of paper, one to a slip, and placed in a bag. If you draw a slip at random, what is the probability it will be a vowel?

Expressing Probabilities

You can express a probability as a fraction, as shown before. But just as you can write a fraction as a decimal, ratio, or percent, you can also write a probability in any of those forms (p. 154).

The probability of getting an even number when you roll a 1–6 number cube is $\frac{1}{2}$. You can also express the probability as follows:

Fraction	Decimal	Ratio	Percent
$\frac{1}{2}$	0.5	1:2	50%

Check It Out

Express each of the following probabilities as a fraction, decimal, ratio, and percent.

8. the probability of drawing a spade when drawing a card from a deck of cards

9. the probability of getting a green counter when drawing a counter from a bag containing 4 green counters and 6 white ones

Strip Graphs

When you conduct an experiment, such as tossing a coin, you need to find a way to show the outcome of each toss. One way to show the outcomes of an experiment is to use a **strip graph**.

The following strip graph shows the results of spinning a spinner.

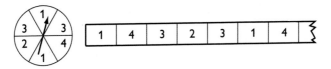

The strip graph shows that the following results occurred on the first seven tosses: 1, 4, 3, 2, 3, 1, 4.

To make a strip graph:
• Draw a series of boxes in a long strip.
• Enter each outcome in a box.

4·6 PROBABILITY

Check It Out
Consider the following strip graph.

| H | T | H | H | H | T |

10. Describe the first six outcomes.
11. Make a strip graph to show the outcomes as you roll a 1–6 number cube ten times. Compare your graphs to others' graphs.

Lottery Fever

**NO WINNER IN WEEKS!
PICK-6 LOTTERY NOW!
WORTH 32 MILLION $$$!**

You read the headline. You say to yourself, "Somebody's *bound* to win this time." But the truth is, you would be wrong! The chances of winning a Pick-6 lottery are always the same, and very, very, very small.

The chances of winning a 6-out-of-50 lottery are 1 in 15,890,700 or 1 in about 16 million. For comparison, think about the chances that you will get struck by lightning—a rare occurrence. It is estimated that in the U.S. roughly 260 people are struck by lightning each year. Suppose the population of the U.S. is about 260 million. Would you be more likely to win the lottery or be struck by lightning? See Hot Solutions for answer.

4·6 PROBABILITY

Outcome Grids

You have seen how to use a tree diagram to show possible outcomes. Another way to show the outcomes in an experiment is to use an **outcome grid.** The following outcome grid shows the outcomes when tossing a coin two times:

	Head	Tail
Head	head, head	head, tail
Tail	tail, head	tail, tail

You can use the grid to find the number of ways the coins can come up the same, which is two.

MAKING OUTCOME GRIDS

Make an outcome grid to show the possible results of rolling two 1–6 number cubes, adding the two numbers together.

- List the outcomes of the first type down the side. List the outcomes of the second type across the top.
- Fill in the outcomes.

	1	2	3	4	5	6
1	2	3	4	5	6	7
2	3	4	5	6	7	8
3	4	5	6	7	8	9
4	5	6	7	8	9	10
5	6	7	8	9	10	11
6	7	8	9	10	11	12

Once you have completed the outcome grid, it is easy to count target outcomes and determine probabilities.

Check It Out

12. Make an outcome grid to show the outcomes when making a two-digit number by spinning the spinner twice.

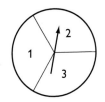

13. What is the probability of getting a number divisible by two when you spin the spinner in item 12 twice?

The WorldPOPClock

5,825,618,337

The U.S. Bureau of the Census estimates how many people are in the world each second on their WorldPOPClock. The estimate is based on projected births and deaths around the world.

At 2 A.M. EST on March 2, 1997, the WorldPOPClock estimate was 5,825,618,337. Using the table below, calculate the world population as of the date you read this page.

Time Unit	Projected Increase
Year	79,178,194
Month	6,598,183
Day	216,927
Hour	9,039
Minute	151
Second	2.5

You can check your answer on the WorldPOPClock on the Internet. Go to http://www.census.gov/ipc/www/popwnote.html then click the link to WorldPOPClock.

Probability Line

You know that the probability of an event is a number from 0 to 1. One way to show probabilities and how they relate to each other is to use a **probability line**. The following probability line shows the possible ranges of probability values:

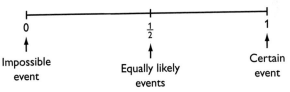

The line shows that events that are certain to happen have a probability of 1. Such an event is the probability of getting a number between 0 and 7 when rolling a standard number cube. An event that cannot happen has a probability of 0. The probability of getting an 8 when spinning a spinner that shows 0, 2, and 4 is 0. Events that are equally likely, such as getting a head or a tail when you toss a coin, have a probability of $\frac{1}{2}$.

Use a number line to show the probabilities of getting a black marble and a blue marble if you draw a marble from a bag of 8 white marbles and 12 black marbles .

SHOWING PROBABILITY ON A PROBABILITY LINE

- Draw a number line and label it from 0 to 1.
- Calculate the probabilities of the given events and show them on the probability line.

   The probability of getting a black marble is $\frac{12}{20} = \frac{3}{5}$ and the probability of getting a blue marble is zero. The probabilities are shown on the following probability line:

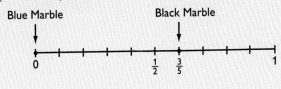

Check It Out

Draw a probability line. Then plot the following:

14. the probability of drawing a white token from a bag of white tokens
15. the probability of tossing two heads on two tosses of a coin

Dependent and Independent Events

If you roll a number cube twice, the result of one does not affect the other. We call these events **independent events.** To find the probability of getting a 4 and then a 6, you can find the probability of each event and then multiply. The probability of getting a 4 on a roll of the number cube is $\frac{1}{6}$ and the probability of getting a 6 is $\frac{1}{6}$. So the probability of getting a 4 followed by a 6 is $\frac{1}{6} \times \frac{1}{6} = \frac{1}{36}$.

Suppose you have six biographies and four novels in a bag. The probability that you get a novel if you choose a book at random is $\frac{4}{10} = \frac{2}{5}$. Once you have taken a novel out, however, there are only nine books left, three of which are novels. So the probability that a friend gets a novel once you have drawn one out is $\frac{3}{9} = \frac{1}{3}$. These events are called **dependent events,** because the probability of one depends on the other.

In the case of dependent events, you still multiply to get the probability of both events happening. So the probability that your friend gets a novel and you also get one is $\frac{2}{5} \times \frac{1}{3} = \frac{2}{15}$.

To find the probability of independent or dependent events:
• Find the probability of the first event.
• Find the probability of the second event.
• Find the product of the two probabilities.

Check It Out

16. Find the probability of getting a head and a 4 if you toss a coin and roll a number cube. Are the events dependent or independent?

17. You draw two balls from a bag containing 9 white tennis balls and 11 yellow ones. What is the probability that you get 2 white tennis balls? Are the events dependent or independent?

Sampling With and Without Replacement

If you draw a card from a deck of cards, the probability that it is a red card is $\frac{26}{52}$, or $\frac{1}{2}$. If you put the card back in the deck and draw another card, the probability that it is a red card is still $\frac{1}{2}$ and the events are independent. This is called **sampling with replacement.**

If you do not put the card back in, the probability of drawing a red card the second time will depend on what you drew the first time. If you drew a red card, there will be only 25 red cards left out of 51 cards, so the probability of drawing a second red card will be $\frac{25}{51}$. In sampling without replacement, the events are dependent.

Check It Out

18. You draw a card from a bag containing cards with the numbers 1–20. You put it back and draw another card. What is the probability that you get two even numbers?

19. What is the answer to item 18 if you do not replace the card?

4•6 PROBABILITY

How Mighty Is the Mississippi?

The legendary Mississippi is the longest river in the United States, but not in the world. Here's how it compares to the world's 12 longest rivers.

River	Location	Length (miles)
Nile	Africa	4,145
Amazon	South America	4,000
Yangtze	Asia	3,915
Yellow	Asia	2,903
Congo	Africa	2,900
Irtysh	Asia	2,640
Mekong	Asia	2,600
Niger	Africa	2,600
Yenisey	Asia	2,543
Parana	South America	2,485
Mississippi	North America	2,348
Missouri	North America	2,315

In this set of data, what is the mean length, the median length, and the range? See Hot Solutions for answer.

 **EXERCISES**

You spin a spinner divided into eight equal parts numbered 1–8.
Find each probability as a fraction, decimal, ratio, and percent.

1. *P*(odd number)

2. *P*(1 or 2)

3. If you choose a marble from a bag of marbles containing 4 red and 6 black marbles, what is the probability of getting a red marble? Is this experimental or theoretical probability?

4. If you toss a thumbtack 50 times and it lands up 15 times, what is the probability of it landing up? Is this experimental or theoretical probability?

5. Draw a probability line to show the probability of getting a number less than 7 when rolling a number cube.

6. Make a strip graph to show the following outcomes when spinning a spinner: red, blue, green, yellow, red, green, green, blue.

7. Make an outcome grid to show the outcomes of spinning a spinner containing the numbers 1–4 and tossing a coin.

In a bag, place 11 pieces of paper each containing one of the letters of the word *Mississippi*.

8. Find the probability of drawing two vowels from the bag if you do not replace the letters between drawings.

9. Find the probability of drawing two vowels if you replace the letters between drawings.

10. In which of items 8 and 9 are the events independent?

What have you learned?

You can use the problems and the list of words that follow to see what you have learned in this chapter. To find out more about a particular problem or word, refer to the boldfaced topic number (for example, **4•2**).

Problem Set

1. Laila listed the businesses in her community on slips of paper and then drew 20 names to choose businesses to survey. Is this a random sample? **4•1**

2. A survey asked, "Do you care enough about the environment to recycle?" Is this question biased? If so, rewrite it. **4•1**

3. How are a frequency graph and a histogram alike? How are they different? **4•2**

4. On a circle graph, the number of degrees to show one part is 180. What percent of the circle is that? **4•2**

Use the following line graph to answer items 5–7, which shows the average low temperatures each month in Seattle. **4•2**

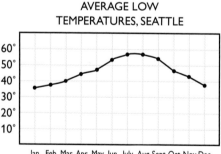

AVERAGE LOW
TEMPERATURES, SEATTLE

Jan Feb Mar Apr May Jun July Aug Sept Oct Nov Dec

5. During which months is the average low less than 40°?
6. What is the range of low temperatures?
7. Is the low in Seattle ever below 0°?

8. On a box plot, the first quartile is 45 and the third quartile is 62. What percent of the data lies between these two figures? **4•2**

9. Is the following statement true or false? Explain. A bar graph shows how a whole is divided. **4•2**

10. Find the mean, median, mode, and range of the numbers 45, 35, 43, 26, and 21. **4•4**

11. Must the median be a member of the set of data? **4•4**

12. $C(6, 4) = ?$ **4•5**   13. $P(6, 4) = ?$ **4•5**

14. What is the probability of rolling a sum of seven when you roll two number cubes? **4•6**

15. A card is drawn from a deck of ten cards, each containing one of the letters of the word *statistics*. A second card is drawn without replacing the first. What is the probability that both cards are *s*? **4•6**

hot **words**

WRITE DEFINITIONS FOR THE FOLLOWING WORDS.

average **4•4**
box plot **4•2**
circle graph **4•2**
combination **4•5**
dependent
  events **4•6**
distribution **4•3**
double-bar
  graph **4•2**
event **4•6**
experimental
  probability **4•6**
factorial **4•5**
histogram **4•2**
independent
  events **4•6**

leaf **4•2**
line graph **4•2**
mean **4•4**
median **4•4**
mode **4•4**
normal
  distribution **4•3**
outcome **4•6**
outcome grid **4•6**
percent **4•2**
permutation **4•5**
population **4•1**
probability **4•6**
probability
  line **4•6**
random
  sample **4•1**
range **4•4**

sample **4•1**
sampling with
  replacement **4•6**
spinner **4•5**
stem **4•2**
stem-and-leaf
  plot **4•2**
strip graph **4•6**
survey **4•1**
table **4•1**
tally marks **4•1**
theoretical
  probability **4•6**
tree diagram **4•5**

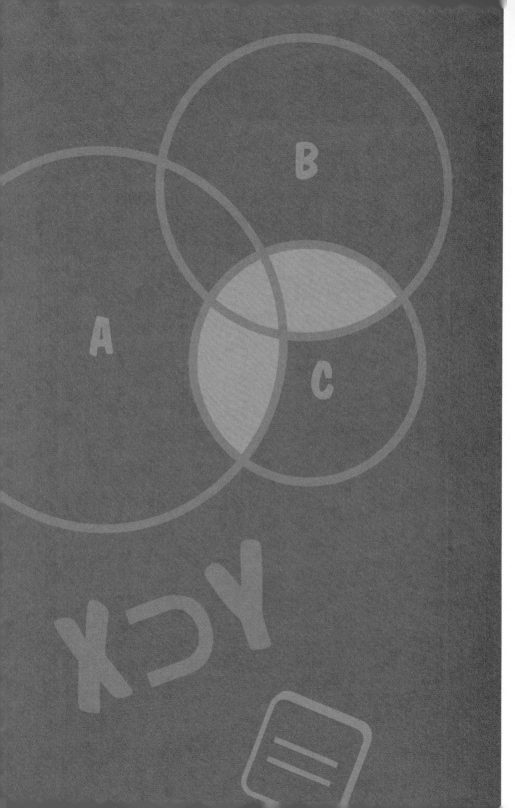

Logic

What do you already know?

You can use the problems and list of words below to see what you already know about this chapter. The answers to the problems are in Hot Solutions at the back of the book, and the definitions of the words are in Hot Words at the front of the book. You can find out more about a particular problem or word by referring to the boldfaced topic number (for example, **5•2**).

Problem Set

Tell whether each statement is true or false.
1. If a conditional statement is true, then its related inverse is always true. **5•1**
2. If a conditional statement is true, then its related contrapositive is true. **5•1**
3. The negation of "It's 7 o'clock" is "It's not 7 o'clock." **5•1**
4. You use a counterexample to show that a statement is true. **5•2**
5. A set is a subset of itself. **5•3**

Write each conditional in if/then form. **5•1**
6. A right angle has a measure of 90°.
7. An acute triangle has three acute angles.

Write the converse of each conditional statement. **5•1**
8. If $n = 8$, then $2 \times n = 16$.
9. If it is summertime, then I wear my white shoes.

Write the negation of each statement. **5•1**
10. Adam got the highest score on the math test.
11. This triangle is not equilateral.

Write the inverse of each conditional statement. **5•1**
12 If you study, then you will receive a good grade.
13. If $a + 1 = 6$, then $a = 5$.

Write the contrapositive of each conditional statement. **5•1**

14. If you are over 12 years old, then you pay the full admission price.

15. If a figure is a triangle, then its area formula is $A = (\frac{1}{2})bh$.

Find a counterexample that shows that each of these statements is false. **5•2**

16. Every month has at least 30 days.

17. 7 and 4 are the only proper factors of 28.

Find all the subsets of each set. **5•3**

18. $\{2, 4\}$

19. $\{2, 4, 6\}$

Find the union of each pair of sets. **5•3**

20. $\{5, 7, 9, 11\} \cup \{7, 11\}$

21. $\{a, b, c, d, e\} \cup \{a, c, e, g\}$

22. $\varnothing \cup \{2, 9\}$

23. $\{x, y\} \cup \{z\}$

24. $\{3, 5, 7\} \cup \{4, 6, 8\}$

Find the intersection of each pair of sets. **5•3**

25. $\{8, 16, 24, 32\} \cap \{2, 4, 8, 16\}$

26. $\{a, c, e\} \cap \{b, d, f\}$

Use the Venn diagram to answer items 27–30. **5•3**

27. List the elements in set A.

28. List the elements in set B.

29. Find $A \cup B$.

30. Find $A \cap B$.

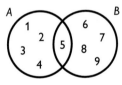

CHAPTER 5

hot **words**

contrapositive **5•1**
converse **5•1**

counterexample **5•2**
intersection **5•3**
inverse **5•1**
union **5•3**
Venn diagram **5•3**

5·1 If/Then Statements

Conditional Statements

A *conditional* is a statement that you can express in if/then form. Often you can rewrite a statement that contains two related ideas as a conditional in if/then form. Do this by placing one of the ideas in the *if* part and the other idea in the *then* part.

Statement: A person who is 21 years old can vote.
The conditional in if/then form:

If a person is 21 years old, then that person can vote.
‌   *if* idea   *then* idea

FORMING CONDITIONAL STATEMENTS

Write this conditional in if/then form:

The people on that basketball team are all over 6 ft tall.

- Find the two ideas.

  (1) people on that basketball team

  (2) people over 6 ft tall

- Decide which idea to put in the *if* part and which to put in the *then* part.

  Idea for the *if* part: people on that basketball team

  Idea for the *then* part: people over 6 ft tall

- Place the ideas in the *if* and *then* parts of the conditional. If necessary, add or change some words so that your sentence makes sense.

  If people are on that basketball team, then those people are over 6 ft tall.

Check It Out

Write each conditional statement in if/then form.
1. A whole number that ends with a 2 is even.
2. A polygon with 3 sides is a triangle.

Converse of a Conditional

When you switch the *if* idea and the *then* idea in a conditional statement, you form a new statement called the **converse.**

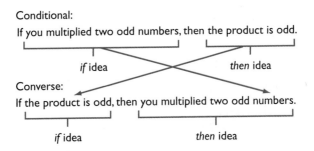

Conditional:
If you multiplied two odd numbers, then the product is odd.

if idea   *then* idea

Converse:
If the product is odd, then you multiplied two odd numbers.

if idea   *then* idea

The converse of a conditional may or may not have the same truth value as the conditional on which it is based.

Check It Out

Write the converse of each conditional.
3. If you multiplied 3 and 4, then you got a product of 12.
4. If an angle has a measure of 90°, then it is a right angle.

Negations and the Inverse of a Conditional

A *negation* of a given statement has the opposite truth value of the given statement. That means that if the given statement is true, the negation is false; if the given statement is false, the negation is true.

Statement: Four is an even number. (True)
Negation: Four is not an even number. (False)

Statement: A triangle has five sides. (False)
Negation: A triangle does not have five sides. (True)

When you negate the *if* idea and the *then* idea of a conditional, you form a new statement called the **inverse.**

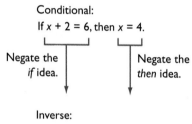

Conditional:
If x + 2 = 6, then x = 4.

Negate the
if idea.

Negate the
then idea.

Inverse:
If x + 2 ≠ 6, then x ≠ 4.

The inverse of a conditional may or may not have the same truth value as the conditional.

Check It Out

Write the negation of each statement.
5. We will go on the class trip.
6. 3 is less than 4.

Write the inverse of each conditional.
7. If an integer ends with 0, then you can divide it by 10.
8. If today is Tuesday, then tomorrow is Wednesday.

Contrapositive of a Conditional

You form the **contrapositive** of a conditional when you negate the *if* idea and the *then* idea, and switch them.

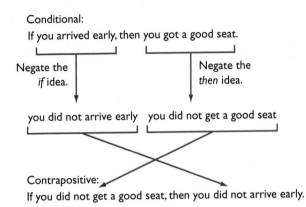

Conditional:
If you arrived early, then you got a good seat.

Negate the *if* idea.

Negate the *then* idea.

you did not arrive early   you did not get a good seat

Contrapositive:
If you did not get a good seat, then you did not arrive early.

Check It Out

Write the contrapositive of each conditional.

9. If you multiplied 6 and 7, then you got a product of 42.

10. If two lines are parallel, then they do not cross.

Who's Who?

That's Tanya sitting in the second seat

I'm Leslie

that's Sylvia sitting in the middle

Suppose you move into a new neighborhood and don't know any of your new classmates. A friend that lives on your block tells you that Tanya, Sylvia, and Leslie sit one behind the other in the first row in math class. She mentions that Tanya is the best math student and always tells the truth, but Sylvia is the worst math student and never tells the truth. You learn from another friend, who lives around the corner, that Leslie sometimes lies and sometimes tells the truth.

When you get to math class, you introduce yourself to the three girls in the first, second, and third seats of the first row. See the picture to find out what they said.

Use logic to decide which girl is Tanya and which is Sylvia. Then decide whether Leslie is lying or telling the truth. See Hot Solutions for answer.

5·1 EXERCISES

Write each conditional in if/then form.
1. An equilateral triangle has three equal sides.
2. An acute angle is smaller than a right angle.
3. Every person who enters the contest wins a prize.
4. An even number is divisible by 2.
5. You add 5 and 6 to get a sum of 11.
6. Yeaphana cleans her room every Saturday.

Write the converse of each conditional.
7. If $x + 3 = 8$, then $x = 5$.
8. If it snows, then it is cold.
9. If a number is divisible by 6, then it is divisible by 3.
10. If this month is May, then next month is June.

Write the negation of each statement.
11. 30 is a multiple of 6.
12. Segment AB is 6 cm long.
13. Saturday is Maria's favorite day of the week.

Write the inverse of each conditional.
14. If $3x = 15$, then $x = 5$.
15. If Sasha misses the bus, then he is late for school.

Write the contrapositive of each conditional.
16. If $x = 4$, then $x + x = 8$.
17. If you subtracted 5 from 12, then you got 7 for an answer.

For each conditional, write the converse, inverse, and contrapositive.
18. If a rectangle has a length of 3 ft and a width of 2 ft, then it has an area of 6 ft^2.
19. If you took good notes, then you passed the course.
20. If it is snowing, then school is canceled.

5·2 Counterexamples

Counterexamples

Any if/then statement is either true or false. One way to decide whether a statement is false is to find just one example that agrees with the *if* idea but not with the *then* idea. Such an example is a **counterexample.**

When reading the conditional below, you may be tempted to think that it is true.

   If a number is prime, then it is an odd number.

The statement is false, however, because there is a counterexample—the number 2. The number 2 agrees with the *if* idea (2 is prime), but it does not agree with the *then* idea (2 is not an odd number.)

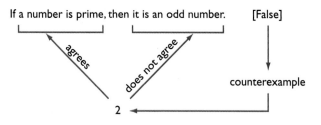

Check It Out

Tell if each statement and its converse are true or false. If a statement is false, give a counterexample.

1. Statement: If a number is greater than 10, then it is greater than 5.

   Converse: If a number is greater than 5, then it is greater than 10.

2. Statement: If one number is odd and another number is even, then their sum is odd.

   Converse: If the sum of two numbers is odd, then one of them is odd and the other is even.

5·2 EXERCISES

Find a counterexample that shows that each of these statements is false.
1. If an angle is acute, then it has a measure less than 80°.
2. If a figure has four sides, then the figure is a square.
3. If $x + y = 13$, then $x = 10$ and $y = 3$.
4. If a number is between 1 and 4, then the number is 2.

Tell whether each conditional is true or false. If false, give a counterexample.
5. If a square has a side of 9 in., then the perimeter of the square is 36 in.
6. If you got a product of 20, then you multiplied 2 and 10.
7. If a whole number is between 1 and 4, then it is a prime number.
8. If you roll two number cubes, then the sum that comes up is even.

Tell if each statement and its converse are true or false. If false, give a counterexample.
9. Statement: If two segments have measures of 5 cm, then the segments are congruent.
   Converse: If two segments are congruent, then the segments have measures of 5 cm.
10. Statement: If $4 \times n = 12$, then the n is 3.
    Converse: If n is 3, then $4 \times n = 12$.

Tell if each statement and its inverse are true or false. If false, give a counterexample.
11. Statement: If $n = 2$, then $n + 9 = 11$.
    Inverse: If $n \neq 2$, then $n + 9 \neq 11$.
12. Statement: If you add 3 and 5, then you get a sum of 8.
    Inverse: If you do not add 3 and 5, then you do not get a sum of 8.

Tell if each statement and its contrapositive are both true or false. If false, give counterexamples.
13. Statement: If a square has a perimeter of 28 ft, then its sides are 7 ft.
    Contrapositive: If the sides of a square are not 7 ft, then its perimeter is not 28 ft.
14. Statement: If a number is even, then it is divisible by 4.
    Contrapositive: If a number is not divisible by 4, then it is not even.

15. Write your own conditional, then its converse, inverse, and contrapositive.

 **Sets**

Sets and Subsets

A *set* is a collection of objects. Each object is called a *member* or *element* of the set. Sets are often named with capital letters.

$A = \{1, 2, 3\}$ $B = \{x, y, z\}$

When a set has no elements, it is an *empty set*. You write { } or ∅ to show the empty set.

When all the elements of a set are also elements of another set, the first set is a *subset* of the other set.

$\{1, 3\}$ is a subset of $\{1, 2, 3\}$

$\{1, 3\} \subset \{1, 2, 3\}$ ($\subset$ is the subset symbol.)

Remember that every set is a subset of itself and that the empty set is a subset of every set.

Check It Out

Tell whether each statement is true or false.

1. $\{4\} \subset \{1, 3, 5\}$ 2. $\emptyset \subset \{3, 6\}$

Find all the subsets of each set.

3. $\{4, 8\}$ 4. $\{m\}$

Union of Sets

You find the **union** of two sets by creating a new set with all the elements from the two sets.

$R = \{1, 3, 5\}$ $T = \{2, 4, 6\}$

$R \cup T = \{1, 2, 3, 4, 5, 6\}$ ($\cup$ is the union symbol.)

When the sets have elements in common, list the common elements only once in the intersection.

$P = \{7, 8, 9\}$ $Q = \{6, 9, 12\}$

$P \cup Q = \{6, 7, 8, 9, 12\}$

Check It Out

Find the union of each pair of sets.

5. $\{1, 2\} \cup \{7, 8\}$ 6. $\emptyset \cup \{5, 10\}$

Intersection of Sets

You find the **intersection** of two sets by creating a new set that contains all the elements that are common to both sets.

$$J = \{1, \text{③}, \text{⑤}, 7\}$$

$$S = \{\text{③}, 4, \text{⑤}\}$$

$$J \cap S = \{3, 5\}$$

If the sets have no elements in common, the intersection is the empty set ($\emptyset$).

Check It Out

Find the intersection of each pair of sets.

7. $\{6, 12\} \cap \{12\}$ 8. $\{8, 10\} \cap \{9, 11\}$

Venn Diagrams

A **Venn diagram** shows you how the elements of two or more sets are related.

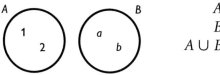

$A = \{1, 2\}$
$B = \{a, b\}$
$A \cup B = \{1, 2, a, b\}$

The separate circles for A and B tell you that the sets have no elements in common. That means that $A \cap B = \emptyset$.

When the circles in a Venn diagram overlap, the overlapping part contains the elements that are common to both sets. This diagram shows some sets of attribute shapes.

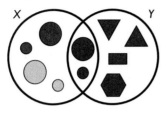

$X = \{$circles$\}$
$Y = \{$blue shapes$\}$

The overlapping parts of X and Y contain shapes that have the attributes of both sets, or $X \cap Y = \{$blue circles$\}$.

With more complex Venn diagrams you have to look carefully to identify the overlapping parts and see which elements of the sets are in those parts. In this diagram A overlaps B, A overlaps C, and B overlaps C. The shaded part of the diagram shows where all three sets overlap one another.

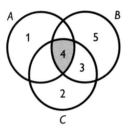

$A = \{1, 4\}$ $A \cup B = \{1, 3, 4, 5\}$
$B = \{3, 4, 5\}$ $A \cup C = \{1, 2, 3, 4\}$
$C = \{2, 3, 4\}$ $B \cup C = \{2, 3, 4, 5\}$

$A \cup B \cup C = \{1, 2, 3, 4, 5\}$

$A \cap B = \{4\}$
$A \cap C = \{4\}$
$B \cap C = \{3, 4\}$

Where all three sets overlap, you can see that $A \cap B \cap C = \{4\}$.

Check It Out

Use this Venn diagram for the following exercises.

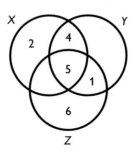

9. List the elements in set Y.
10. Find $X \cap Z$.
11. Find $Y \cup Z$.
12. Find $Y \cap Z$.
13. Find $X \cap Y \cap Z$.

5·3 EXERCISES

Tell whether each statement is true or false.
1. $\{2, 4, 5\} \subset \{\text{odd numbers}\}$
2. $\varnothing \subset \{4, 5\}$
3. $\{4, 8, 12\} \subset \{\text{even numbers}\}$
4. List all the subsets of $\{1, 2, 3\}$

Find the union of each pair of sets.
5. $\{2, 3\} \cup \{7, 8\}$
6. $\{r, s\} \cup \{s, t\}$
7. $\{c, h, u, r, n\} \cup \{b, u, r, n\}$
8. $\{6, 12, 18, 24\} \cup \{6, 24\}$

Find the intersection of each pair of sets.
9. $\{2, 4, 6\} \cap \{3, 6, 9, 12\}$
10. $\{5, 10, 15\} \cap \{6, 12, 18\}$
11. $\{c, h, u, r, n\} \cap \{b, u, r, n\}$
12. $\varnothing \cap \{8\}$

Use the Venn diagram below for items 13–16.

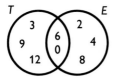

13. List set T.
14. List set E.
15. Find $T \cup E$.
16. Find $T \cap E$.

Use the Venn diagram below for items 17–20.

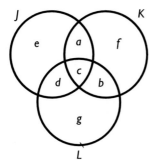

17. List the elements of set J.
18. Find $J \cup K$.
19. Find $L \cap K$.
20. Find $J \cap K \cap L$.

What have you learned?

You can use the problems and list of words below to see what you have learned in this chapter. You can find out more about a particular problem or word by referring to the boldfaced topic number (for example, **5•2**).

Problem Set

Tell whether each statement is true or false.
1. You form the inverse of a conditional by negating the *if* idea and the *then* idea. **5•1**
2. If a conditional statement is false, then its related contrapositive is false. **5•2**
3. The negation of "It's winter" is "It's not winter." **5•1**
4. You use a counterexample to show that a statement is false. **5•2**
5. A counterexample agrees with the *if* idea and the *then* idea of a conditional. **5•2**
6. The empty set is a subset of every set. **5•3**
7. You form the intersection of two sets by making a set of elements that are common to both sets. **5•3**

Write each conditional in if/then form. **5•1**
8. An obtuse triangle has one obtuse angle.
9. Summer is not a good time of year to go skiing.

Write the converse of each conditional statement. **5•1**
10. If $d = 4$, then $d + 5 = 9$.
11. If you are not dressed warmly, then you feel cold.

Write the negation of each statement. **5•1**
12. The store is open late on Thursday.
13. These two lines are not parallel.

Write the inverse of each conditional statement. **5•1**
14. If you eat the right foods, then your health improves.
15. If $b + 2 = 7$, then $b = 5$.

Write the contrapositive of each conditional statement. **5•1**

16. If you are 12 years old, then you qualify for membership in the club.

17. If you returned the library book on time, then you did not pay a fine.

Find a counterexample that shows that each of these statements is false. **5•2**

18. Vermont is the only state that begins with the letter *v*.

19. The number 7 has only odd multiples.

Find all the subsets of each set. **5•3**

20. {5, 7}

21. {5, 7, 9}

Find the union of each pair of sets. **5•3**

22. {r, s, t, v} ∪ {r, t, w}

23. {0} ∪ {2, 4}

Find the intersection of each pair of sets. **5•3**

24. {7, 14, 21, 28} ∩ {14, 28, 42}

25. {even numbers} ∩ {odd numbers}

Use the Venn diagram for items 26–30. **5•3**

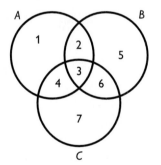

26. List set A.
27. List set C.
28. Find A ∪ B.
29. Find B ∩ C.
30. Find A ∩ B ∩ C.

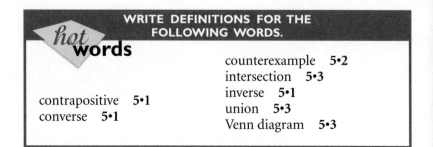

WRITE DEFINITIONS FOR THE FOLLOWING WORDS.

hot **words**

contrapositive **5•1**
converse **5•1**

counterexample **5•2**
intersection **5•3**
inverse **5•1**
union **5•3**
Venn diagram **5•3**

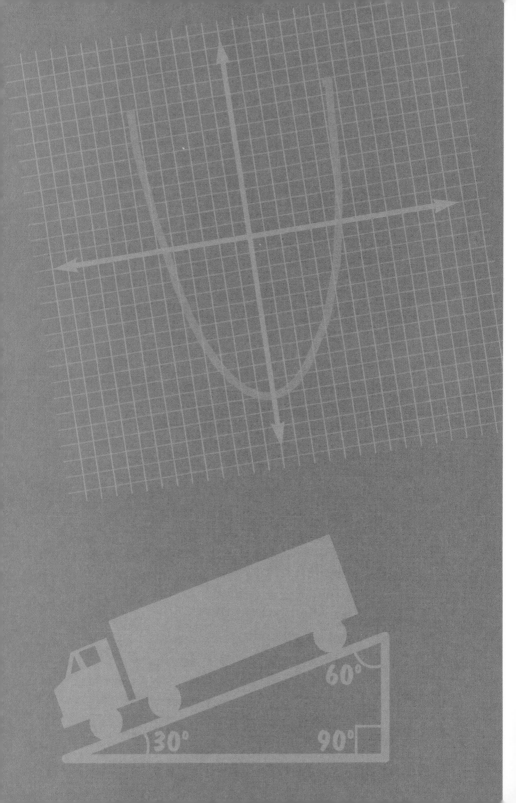

Algebra

Problem Set

Write an expression for each phrase. **6•1**
1. a number increased by 5
2. the product of 4 and some number
3. twice the difference of a number and 6
4. two less than the quotient of a number and 3

Write an equation for each sentence. **6•1**
5. If 3 is subtracted from twice a number, the result is 5 more than the number.
6. 6 times the sum of a number and 2 is 4 less than twice the number.

Factor out the greatest common factor in each expression. **6•2**
7. $6x + 18$
8. $10n - 15$
9. $3a - 21$

Simplify each expression. **6•2**
10. $8x + 7 - 5x$
11. $9a + 7b - a - 5b$
12. $4(2n - 1) - (n + 6)$

13. Find the distance traveled by a jogger who jogs at 6 mi/hr for $2\frac{1}{2}$ hr. Use the formula $d = rt$. **6•3**

Use a proportion to solve each problem. **6•4**
14. In a class, the ratio of boys to girls is $\frac{2}{3}$. If there are 15 girls in the class, how many boys are there?
15. A map is drawn using a scale of 80 km to 1 cm. The distance between two cities is 600 km. How far apart are the two cities on the map?

Solve each inequality. Graph the solution. **6•5**
16. $x + 4 < 7$
17. $3x \geq 12$
18. $n - 8 > -6$
19. $\frac{n}{2} \leq -1$

Locate each point on the coordinate plane and tell where it lies. **6•6**
20. $A(1, 3)$
21. $B(-3, 0)$
22. $C(3, -2)$
23. $D(0, -4)$
24. $E(-2, 5)$
25. $F(-4, -3)$

Find five solutions of each equation. Graph each line. **6•6**
26. $y = 3x - 5$
27. $y = -x + 2$
28. $y = \frac{1}{2}x - 1$

Graph each line. **6•6**
29. $x = -2$
30. $y = 3$

CHAPTER 6

hot **words**

associative property **6•2**
axes **6•6**
commutative property **6•2**
cross product **6•4**
difference **6•1**
distributive property **6•2**
equation **6•1**
equivalent **6•1**

equivalent expression **6•2**
expression **6•1**
formula **6•3**
horizontal **6•6**
inequality **6•5**
like terms **6•2**
order of operations **6•3**
ordered pair **6•6**
origin **6•6**
perimeter **6•3**
point **6•6**
product **6•1**

proportion **6•4**
quadrant **6•6**
quotient **6•1**
rate **6•4**
ratio **6•4**
solution **6•6**
sum **6•1**
term **6•1**
variable **6•1**
vertical **6•6**
x-axis **6•6**
y-axis **6•6**

WHAT DO YOU KNOW?

6·1 Writing Expressions and Equations

Expressions

In mathematics, often the value of a particular number may be unknown. A **variable** is a symbol, usually a letter, that is used to represent an unknown number. Here are some commonly used variables:

$$x \qquad n \qquad y \qquad a \qquad ?$$

A **term** can be a number, a variable, or a number and variable combined by multiplication or division. Some examples of terms are:

$$w \qquad 5 \qquad 3x \qquad \frac{y}{8}$$

An **expression** can be a term or a collection of terms separated by addition or subtraction signs. The table below shows some expressions with the number of terms listed.

Expression	Number of Terms	Description
$5y$	1	A number is multiplied by a variable.
$6z + 4$	2	Terms are separated by a +.
$3x + 7a - 5$	3	
$\dfrac{9xz}{y}$	1	All multiplication and division; no + symbol.

Check It Out

Count the number of terms in each expression.

1. $6n + 4$
2. $7bc$
3. $5m - 3n - 4$
4. $2(x - 3) + 10$

Writing Expressions Involving Addition

To write an expression, you often have to interpret a written phrase. For example, the phrase "4 added to some number" can be written as the expression $x + 4$, where the variable x represents the unknown number.

Notice that the words "added to" indicate that the operation between 4 and the number is to be addition. Other words and phrases that indicate addition are "more than," "plus," and "increased by." One other word that indicates addition is **sum.** The sum of two terms is the result of adding them together.

The following table shows some common phrases and their corresponding expressions.

Phrase	Expression
3 more than some number	$n + 3$
a number increased by 7	$x + 7$
9 plus some number	$9 + y$
the sum of a number and 6	$n + 6$

Check It Out

Write an expression for each phrase.
5. a number added to 5
6. the sum of a number and 4
7. some number increased by 8
8. 2 more than some number

Writing Expressions Involving Subtraction

The phrase "4 subtracted from some number" can be written as the expression $x - 4$, where the variable x represents the unknown number. Notice that the words "subtracted from" indicate that the operation between the number and 4 is to be subtraction.

Some other words and phrases that indicate subtraction are "less than," "minus," and "decreased by." One other word that indicates subtraction is **difference.** The difference between two terms is the result of subtracting them.

In a subtraction expression, the order of the terms is very important. You have to know which term is being subtracted and which is being subtracted from. To help interpret the phrase "6 less than a number," replace "a number" with 10. What is 6 less than 10? The answer is 4, which is $10 - 6$, not $6 - 10$. The phrase translates to the expression $x - 6$, not $6 - x$.

Some common phrases and their corresponding expressions are listed in the table below.

Phrase	Expression
5 less than some number	$n - 5$
a number decreased by 8	$x - 8$
7 minus some number	$7 - y$
the difference between a number and 2	$n - 2$

Check It Out

Write an expression for each phrase.

9. a number subtracted from 8
10. the difference between a number and 3
11. some number decreased by 6
12. 4 less than some number

Writing Expressions Involving Multiplication

The phrase "4 multiplied by some number" can be written as the expression $4x$, where the variable x represents the unknown number. Notice that the words "multiplied by" indicate that the operation between the number and 4 is to be multiplication.

Some other words and phrases that indicate multiplication are "times," "twice," and "of." "Twice" is used to mean "2 times." "Of" is used primarily with fractions and percents. One other word that indicates multiplication is **product.** The product of two terms is the result of them being multiplied.

Here are some common phrases and their corresponding expressions.

Phrase	Expression
5 times some number	$5a$
twice a number	$2x$
one-fourth of some number	$\frac{1}{4}y$
the product of a number and 8	$8n$

Check It Out
Write an expression for each phrase.
13. a number multiplied by 4
14. the product of a number and 8
15. 25% of some number
16. 9 times some number

Writing Expressions Involving Division

The phrase "4 divided by some number" can be written as the expression $\frac{4}{x}$, where the variable x represents the unknown number. Notice that the words "divided by" indicate that the operation between the number and 4 is to be division.

Some other words and phrases that indicate division are "ratio of" and "divide." One other word that indicates division is **quotient.** The quotient of two terms is the result of one being divided by the other.

Some common phrases and their corresponding expressions appear in the table.

Phrase	Expression
the quotient of 20 and some number	$\dfrac{20}{n}$
a number divided by 6	$\dfrac{x}{6}$
the ratio of 10 and some number	$\dfrac{10}{y}$
the quotient of a number and 5	$\dfrac{n}{5}$

Check It Out

Write an expression for each phrase.
17. a number divided by 5
18. the quotient of 8 and a number
19. the ratio of 20 and some number
20. the quotient of some number and 4

Writing Expressions Involving Two Operations

To translate the phrase "4 added to the product of 5 and some number" to an expression, first realize that "4 added to" means "something" + 4. That "something" is "the product of 5 and some number," which is $5x$, because "product" indicates multiplication. Thus the expression can be written as $5x + 4$.

Phrase	Expression	Think
2 less than the quotient of a number and 5	$\dfrac{x}{5} - 2$	"2 less than" means "something" − 2; "quotient" indicates division.
5 times the sum of a number and 3	$5(x + 3)$	Write the sum inside parentheses so that the entire sum is multiplied by 5.
3 more than 7 times a number	$7x + 3$	"3 more than" means "something" + 3; "times" indicates multiplication.

Check It Out

Translate each phrase to an expression.

21. 6 less than the product of 4 and a number
22. 5 subtracted from the quotient of 8 and a number
23. twice the difference between a number and 4

Three Astronauts and a Cat

Here is a modern version of a problem that first appeared in the year 850.

Three astronauts and their pet cat land on a deserted asteroid that resembles Earth in many ways. They find a large lake with lots of fish in it, and they try to catch as many fish as they can before nightfall. Tired, they take shelter and decide to divide up the fish in the morning.

One astronaut wakes up during the night and decides to take her share. She divides the pile of fish into three equal parts, but there is one left over. So she gives it to the cat. She hides her third and puts the rest of the fish back in a pile. Later the second and third astronauts wake up in turn and do exactly the same thing. In the morning they divide the pile of fish that's left into three equal parts. They give the one remaining fish to the cat. What is the smallest number of fish they originally caught? See Hot Solutions for answer.

6•1 WRITING EXPRESSIONS

Equations

An expression is a phrase; an **equation** is a sentence. An equation indicates that two expressions are **equivalent,** or equal. The symbol used in an equation is the equals sign, $=$.

To translate the sentence "2 less than the product of a number and 5 is the same as 6 more than the number" to an equation, first identify the words that indicate "equals." In this sentence, "equals" is indicated by "is the same as." In other sentences "equals" may be "is," "the result is," "you get," or just "equals."

Once you have identified the $=$, you can then translate the phrase that comes before the $=$ and write the expression on the left side. Then translate the phrase that comes after the $=$ and write the expression on the right side.

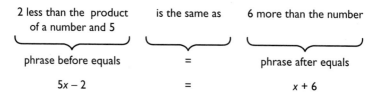

2 less than the product of a number and 5	is the same as	6 more than the number
phrase before equals	$=$	phrase after equals
$5x - 2$	$=$	$x + 6$

Check It Out

Write an equation for each sentence.
24. 9 subtracted from 5 times a number is 6.
25. If 6 is added to the quotient of a number and 3, the result is 2 less than the number.
26. 1 less than 4 times a number is twice the sum of the number and 5.

 **6·1 EXERCISES**

Count the number of terms in each expression.

1. $8x + 2$ 2. 5

3. $6x - 3y + 5z$ 4. $2(n - 4) - 3$

Write an expression for each phrase.

5. 2 more than a number   6. a number added to 6

7. the sum of a number and 4   8. 5 less than a number

9. 12 decreased by some number

10. the difference between a number and 3

11. one third of some number

12. twice a number

13. the product of a number and 8

14. a number divided by 7

15. the ratio of 10 and some number

16. the quotient of a number and 6

Write an expression for each phrase.

17. 9 more than the product of a number and 3

18. 3 less than twice a number

19. twice the sum of 6 and a number

Write an equation for each sentence.

20. 2 more than the quotient of a number and 3 is the same as 2 less than the number.

21. If 7 is subtracted from twice a number, the result is 9.

22. 4 times the sum of a number and 5 is 8 more than twice the number.

23. Which of the following words is used to indicate multiplication?
    A. Sum   B. Difference
    C. Product   D. Quotient

24. Which of the following does not indicate subtraction?
    A. Less than   B. Difference
    C. Decreased by   D. Ratio of

25. Which of the following shows "twice the sum of a number and 4"?
    A. $2(x + 4)$ B. $2x + 4$
    C. $2(x - 4)$ D. $2 + (x + 4)$

6·1 EXERCISES

6·2 Simplifying Expressions

Terms

As you may remember, terms can be numbers, variables, or numbers and variables combined by multiplication or division. Some examples of terms are:

$$n \qquad 7 \qquad 5x \qquad x^2 \qquad 3(n + 5)$$

Compare the terms 7 and $5x$. The value of $5x$ will change as the value of x changes. If $x = 2$, then $5x = 5(2) = 10$, and if $x = 3$, then $5x = 5(3) = 15$. Notice, though, that the value of 7 never changes—it remains constant. When a term contains a number only, it is called a *constant* term.

✓ Check It Out
Decide whether each term is a constant term.
1. $6x$ 2. 9
3. $3(n + 1)$ 4. 5

The Commutative Property of Addition and Multiplication

The **commutative property** of addition states that the order of two terms being added may be switched without changing the result; $3 + 4 = 4 + 3$ and $x + 8 = 8 + x$. The commutative property of multiplication states that the order of two terms being multiplied may be switched without changing the result; $3(4) = 4(3)$ and $x \cdot 8 = 8x$.

The commutative property does not hold for subtraction or division. The order of the terms does affect the result: $5 - 3 = 2$, but $3 - 5 = -2$; $8 \div 4 = 2$, but $4 \div 8 = \frac{1}{2}$.

✓ Check It Out
Rewrite each expression, using the commutative property of addition or multiplication.
5. $2x + 7$ 6. $n \cdot 6$
7. $5 + 4y$ 8. $3 \cdot 8$

The Associative Property of Addition and Multiplication

The **associative property** of addition states that the grouping of three terms being added does not affect the result: $(3 + 4) + 5 = 3 + (4 + 5)$ and $(x + 6) + 10 = x + (6 + 10)$. The associative property of multiplication states that the grouping of three terms being multiplied does not affect the result: $(2 \cdot 3) \cdot 4 = 2 \cdot (3 \cdot 4)$ and $5 \cdot 3x = (5 \cdot 3)x$.

The associative property does not hold for subtraction or division. The grouping of the numbers does affect the result: $(8 - 6) - 4 = -2$, but $8 - (6 - 4) = 6$; $(16 \div 8) \div 2 = 1$, but $16 \div (8 \div 2) = 4$.

...3, 2, 1, Blast Off

Fleas, those tiny pests that make your dog or cat itch, are amazing jumpers. Actually, fleas don't jump but launch themselves at a rate 50 times faster than the space shuttle leaves Earth.

Elastic pads on the flea's feet compress like a coil spring to power a liftoff. When a flea is ready to leave, it hooks onto its host, locks its legs in place, then releases the "hooks." The pads spring back into shape and the flea blasts off. A flea no bigger than 0.05 of an inch can propel itself a distance of as much as 8 inches—160 times its own length! If you could match a flea's feat, how far could you jump? See Hot Solutions for answer.

Check It Out

Rewrite each expression, using the associative property of addition or multiplication.

9. $(4 + 5) + 8$
10. $(2 \cdot 3) \cdot 5$
11. $(5x + 4y) + 3$
12. $6 \cdot 9n$

The Distributive Property

The **distributive property** of addition and multiplication states that multiplying a sum by a number is the same as multiplying each addend by that number and then adding the two products. So $3 \times (2 + 3) = (3 \times 2) + (3 \times 3)$.

How would you multiply 7×99 in your head? You might think, $700 - 7 = 693$. If you did, you have used the distributive property.

$7(100 - 1)$ • Distribute the factor of 7 to each term inside the parentheses.

$= (7 \cdot 100) - (7 \cdot 1)$ • Simplify, using order of operations.

$= 700 - 7$

$= 693$

The distributive property does not hold for division.

$3 \div (2 + 3) \neq (3 \div 2) + (3 \div 3).$

Check It Out

Use the distributive property to find each product.

13. $6 \cdot 99$
14. $3 \cdot 106$
15. $4 \cdot 198$
16. $5 \cdot 211$

Equivalent Expressions

The distributive property can be used to write an **equivalent expression** with two terms. Equivalent expressions are two different ways of writing one expression.

WRITING AN EQUIVALENT EXPRESSION

Write an equivalent expression for $5(9x - 7)$.

$5(9x - 7)$ • Distribute the factor of 5 to each term inside the parentheses.

$5 \cdot 9x - 5 \cdot 7$

$45x - 35$ • Simplify.

$5(9x - 7) = 45x - 35$ • Write the equivalent expressions.

Distributing When the Factor Is Negative

The distributive property is applied in the very same way if the factor to be distributed is negative.

Write an equivalent expression for $-3(5x - 6)$.

$-3(5x - 6)$ • Distribute the factor of –3 to each item inside the parentheses.

$(-3 \cdot 5x) - (-3 \cdot 6)$ • Simplify.

$-15x - (-18)$

$-15x + 18$

$-3(5x - 6) = -15x + 18$ • Write the equivalent expressions.

Check It Out

Write an equivalent expression.

17. $2(3x + 1)$
18. $6(2n - 3)$
19. $-1(8y - 2)$
20. $-2(-5x + 4)$

The Distributive Property
with Common Factors

Given the expression $10x + 15$, you can use the distributive property to write an equivalent expression. Recognize that each of the two terms has a factor of 5.

Rewrite the expression as $5 \cdot 2x + 5 \cdot 3$. Then write the common factor 5 in front of the parentheses and the remaining factors inside the parentheses: $5(2x + 3)$. You have used the distributive property to *factor out the common factor.*

FACTORING OUT THE COMMON FACTOR

Factor out the common factor from the expression $12n - 30$.

$12n - 30$	• Find a common factor.
$(6 \cdot 2n) - (6 \cdot 5)$	• Rewrite the expression.
$6(2n - 5)$	• Use the distributive property.
$12n - 30 = 6 \cdot (2n - 5)$	

Always be sure to factor out the greatest common factor. For example, for $16n-8$, you could factor out 2, $2(8n-4)$, but this would not be fully factored. $8(2n-1)$ has the greatest common factor factored out.

✔ Check It Out

Factor out the greatest common factor in each expression.
21. $7x + 14$
22. $4n - 10$
23. $10c + 50$
24. $18a - 27$

Like Terms

Like terms are terms that contain the same variable with the same exponent. Constant terms are like terms because they do not have any variables. Some examples of like terms are listed below.

Like Terms	Reason
$3x$ and $4x$	Both contain the same variables.
3 and 11	Both are constant terms.
$2n^2$ and $6n^2$	Both contain the same variable with the same exponent.

Some examples of terms that are not like terms are listed below.

Not like Terms	Reason
$3x$ and $5y$	Variables are different.
$4n$ and 12	One term has a variable; the other is constant.
$2x^2$ and $2x$	The variables are the same, but the exponents are different.

Two terms that are like terms may be combined into one term by adding or subtracting. Consider the expression $3x + 4x$. Notice that the two terms have a common factor, x. Use the distributive property to write $x(3 + 4)$. This simplifies to $7x$, so $3x + 4x = 7x$.

COMBINING LIKE TERMS

Simplify $3n - 5n$.

- Recognize that the variable is a common factor. Rewrite the expression, using the distributive property.
$$n(3 - 5)$$
- Simplify. $\quad n(-2)$
- Use the commutative property of multiplication.
$$-2n$$

6·2 SIMPLIFYING EXPRESSIONS

Check It Out

Combine like terms.

25. $5x + 6x$
26. $10y - 4y$
27. $4n + 3n + n$
28. $2a - 8a$

Simplifying Expressions

Expressions are simplified when all of the like terms have been combined. Terms that are not like terms cannot be combined. In the expression $3x - 5y + 6x$, there are three terms. Two of them are like terms, $3x$ and $6x$, which combine to be $9x$. The expression can be written as $9x - 5y$, which is simplified because the two terms are not like terms.

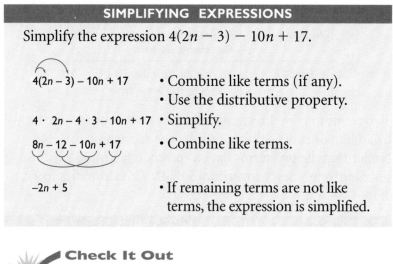

SIMPLIFYING EXPRESSIONS

Simplify the expression $4(2n - 3) - 10n + 17$.

$4(2n - 3) - 10n + 17$ • Combine like terms (if any).
                   • Use the distributive property.

$4 \cdot 2n - 4 \cdot 3 - 10n + 17$ • Simplify.

$8n - 12 - 10n + 17$ • Combine like terms.

$-2n + 5$ • If remaining terms are not like terms, the expression is simplified.

Check It Out

Simplify each expression.

29. $6y + 3z - 4y + z$
30. $2x + 3(x - 2)$
31. $7a + 8 - 2(a + 3)$
32. $4(2n - 1) - (n - 6)$

 **EXERCISES**

Decide whether each term is a constant term.
1. $6n$ 2. -5

Using the commutative property of addition or multiplication, rewrite each expression.
3. $3 + 8$ 4. $n \cdot 5$
5. $4x + 7$

Using the associative property of addition or multiplication, rewrite each expression.
6. $3 + (6 + 8)$ 7. $(4 \cdot 5) \cdot 3$
8. $5 \cdot 2n$

Use the distributive property to find each product.
9. $8 \cdot 99$ 10. $3 \cdot 106$

Write an equivalent expression.
11. $3(4x + 1)$ 12. $-5(2n + 3)$
13. $10(3a - 4)$ 14. $-(-5y - 2)$

Factor out the greatest common factor in each expression.
15. $6x + 12$ 16. $4n - 28$
17. $20a - 30$

Combine like terms.
18. $10x - 3x$ 19. $5n + 4n - n$
20. $4a - 6a$

Simplify each expression.
21. $5a + 3b - a - 4b$ 22. $2x + 2(3x - 4) + 5$
23. $-2(-3n - 1) - (n + 3)$

24. Which property is illustrated by $5(2x + 1) = 10x + 5$?
     A. Commutative property of multiplication
     B. Distributive property
     C. Associative property of multiplication
     D. The example does not illustrate a property.
25. Which of the following shows the expression $24x - 36$ with the greatest common factor factored out?
     A. $2(12x - 18)$ B. $3(8x - 12)$
     C. $6(4x - 6)$ D. $12(2x - 3)$

6·3 Evaluating Expressions and Formulas

Evaluating Expressions

Once an expression has been written, you can *evaluate* it for different values of the variable. To evaluate $5x - 3$ for $x = 4$, *substitute* 4 in place of the x: $5(4) - 3$. Use **order of operations** to evaluate: multiply first, then subtract. So $5(4) - 3 = 20 - 3 = 17$.

EVALUATING AN EXPRESSION

Evaluate $3(x - 2) - \frac{12}{x} + 5$ for $x = 2$.

- Substitute the value for x. $\qquad\qquad$ $3[(2) - 2] - \frac{12}{2} + 5$

- Use order of operations to simplify. $\quad$ $3(0) - \frac{12}{2} + 5$
  Simplify within parentheses, then
  evaluate powers.

- Multiply and divide, in order $\qquad\quad$ $0 - 6 + 5$
  from left to right.

- Add and subtract, in order $\qquad\qquad$ -1
  from left to right.

When $x = 2$, $3(x - 2) - \dfrac{12}{x} + 5 = -1$.

Check It Out

Evaluate each expression for the given value.
1. $4x - 5$ for $x = 4$
2. $2a + 5 - \frac{6}{a}$ for $a = 3$
3. $6(n - 5) - 2n + 3$ for $n = 9$
4. $2(3y - 2) + 3y$ for $y = 2$

Evaluating Formulas

The Formula for Perimeter of a Rectangle

The **perimeter** of a rectangle is the distance around the rectangle. The **formula** $P = 2w + 2l$ can be used to find the perimeter, P, if the width, w, and the length, l, are known.

FINDING THE PERIMETER OF A RECTANGLE

Find the perimeter of a rectangle whose width is 3 ft and length is 4 ft.

- Use the formula for the perimeter of a rectangle ($P = 2w + 2l$).

  $P = 2w + 2\ell$

- Substitute the given values for w and l.

  $= (2 \times 3) + (2 \times 4)$

- Multiply.

  $= 6 + 8$

- Add.

  $= 14$

The perimeter of the rectangle is 14 ft.

Check It Out

Find the perimeter of each rectangle described.

5. $w = 6$ cm, $l = 10$ cm
6. $w = 2.5$ ft, $l = 6.5$ ft

The Formula for Distance Traveled

The distance traveled by a person, vehicle, or object depends on its rate and the amount of time. The formula $d = rt$ can be used to find the distance traveled, d, if the rate, r, and the amount of time, t, are known.

FINDING THE DISTANCE TRAVELED

Find the distance traveled by a bicyclist who averages 20 mi/hr for 4 hr.

- Substitute values into the distance formula ($d = rt$). $\qquad d = 20 \times 4$

- Multiply. $\qquad d = 80$

The bicyclist rode 80 miles.

Check It Out

Find the distance traveled, if
7. a person rides 10 mi/hr for 3 hr.
8. a plane flies 600 km/hr for 2 hr.
9. a person drives a car 55 mi/hr for 4 hr.
10. a snail moves 2 ft/hr for 3 hr.

 **EXERCISES**

Evaluate each expression for the given value.
1. $3x - 11$ for $x = 9$
2. $3(6 - a) + 7 - 2a$ for $a = 4$
3. $\frac{n}{2} + 3n - 8$ for $n = 6$
4. $3(2y - 1) - \frac{10}{y} + 6$ for $y = 2$

Use the formula $P = 2w + 2l$ to answer items 5–7.
5. Find the perimeter of a rectangle that is 20 ft long and 15 ft wide.
6. Find the perimeter of the rectangle.

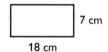

18 cm

7 cm

7. Nia had a 20-in. by 30-in. enlargement made of a photograph. She wanted to have it framed. How many inches of frame would it take to enclose the photo?

Use the formula $d = rt$ to answer items 8–10.
8. Find the distance traveled by a walker who walks at 3 mi/hr for 2 hours.
9. A race car driver averaged 120 mi/hr. If the driver completed the race in $2\frac{1}{2}$ hr, how many miles was the race?
10. The speed of light is approximately 186,000 mi/sec. About how far does light travel in 3 sec?

6·4 Ratio and Proportion

Ratio

A **ratio** is a comparison of two quantities. If there are 10 boys and 15 girls in a class, the ratio of the number of boys to the number of girls is 10 to 15, which can be expressed as the fraction $\frac{10}{15}$, which reduces to $\frac{2}{3}$. You can write some other ratios.

Comparison	Ratio	As a Fraction
Number of girls to number of boys	15 to 10	$\frac{15}{10} = \frac{3}{2}$
Number of boys to number of students	10 to 25	$\frac{10}{25} = \frac{2}{5}$
Number of students to number of girls	25 to 15	$\frac{25}{15} = \frac{5}{3}$

Check It Out

A coin bank contains 8 dimes and 4 quarters. Write each ratio, and reduce to lowest terms.
1. number of quarters to number of dimes
2. number of dimes to number of coins
3. number of coins to number of quarters

Proportions

A **rate** is a ratio that compares a quantity to 1 unit. Some examples of rates are listed below.

$$\frac{55 \text{ mi}}{1 \text{ hr}} \qquad \frac{5 \text{ apples}}{\$1} \qquad \frac{18 \text{ mi}}{1 \text{ gal}} \qquad \frac{\$400}{1 \text{ wk}} \qquad \frac{60 \text{ sec}}{1 \text{ min}}$$

If a car gets $\frac{18 \text{ mi}}{1 \text{ gal}}$, then the car can get $\frac{36 \text{mi}}{2 \text{ gal}}$, $\frac{54 \text{ mi}}{3 \text{ gal}}$, and so on. The ratios are all equal—they can be reduced to $\frac{18}{1}$.

When two ratios are equal, they form a **proportion.** One way to determine whether two ratios form a proportion is to check their **cross products.** Every proportion has two cross products: the numerator of one ratio multiplied by the denominator of the other ratio. If the cross products are equal, the two ratios form a proportion.

DETERMINING A PROPORTION

Determine whether a proportion is formed.

$$\require{cancel}\cancel{\frac{6}{9}}\bcancel{\frac{45}{60}} \qquad \cancel{\frac{15}{9}}\bcancel{\frac{70}{42}}$$ • Find the cross products.

$6 \cdot 60 \stackrel{?}{=} 45 \cdot 9 \qquad 15 \cdot 42 \stackrel{?}{=} 70 \cdot 9$ • If the sides are equal, the ratios are proportional.

$360 \stackrel{?}{=} 405 \qquad\quad 630 \stackrel{?}{=} 630$

$\frac{6}{9} \stackrel{?}{=} \frac{45}{60} \qquad\quad \frac{15}{9} \stackrel{?}{=} \frac{70}{42}$

is not a proportion.   is a proportion.

Check It Out

Determine whether a proportion is formed.

4. $\frac{9}{12} = \frac{15}{20}$

5. $\frac{6}{5} = \frac{20}{17}$

Using Proportions to Solve Problems

To use proportions to solve problems, set up two ratios which relate what you know to what you are solving for.

Suppose that you can buy 5 apples for $2. How much would it cost to buy 17 apples? Let c represent the cost of the 17 apples. If you express each ratio as $\frac{\text{apples}}{\$}$, then one ratio is $\frac{5}{2}$ and another is $\frac{17}{c}$. The two ratios must be equal.

$$\frac{5}{2} = \frac{17}{c}$$

To solve for c, you can use the cross products. Because you have written a proportion, the cross products are equal.

$$5c = 34$$

To isolate the variable, divide by 5 on both sides of the equation and simplify.

$$\frac{5c}{5} = \frac{34}{5} \qquad\qquad c = 6.8$$

So 17 apples would cost $6.80.

> **Check It Out**
> Use proportions to solve items 6 and 7.
> 6. A car gets 20 mi/gal. How many gallons would the car need to travel 70 mi?
> 7. A worker earns $30 every 4 hr. How much would the worker earn in 14 hr?

6·4 EXERCISES

A basketball team has 10 wins and 5 losses. Write each ratio and reduce to lowest terms if possible.

1. number of wins to number of losses

2. number of wins to number of games

3. number of losses to number of games

Determine whether a proportion is formed.

4. $\frac{3}{5} = \frac{7}{11}$ 5. $\frac{9}{6} = \frac{15}{10}$ 6. $\frac{3}{4} = \frac{9}{16}$

Use a proportion to solve each problem.

7. In a class the ratio of boys to girls is $\frac{3}{2}$. If there are 12 boys in the class, how many girls are there?

8. An overseas telephone call costs $0.36 per minute. How much would a 6-minute call cost?

9. A map is drawn using a scale of 40 km to 1 cm. The distance between two cities is 300 km. How far apart are the two cities on the map?

10. A blueprint of a house is drawn using a scale of 5 m to 2 cm. On the blueprint, a room is 6 cm long. How long will the actual room be?

 Inequalities

Showing Inequalities

When comparing the numbers 7 and 4, you might say that "7 is greater than 4," or you might also say "4 is less than 7." When two expressions are not equal, or could be equal, you can write an **inequality.** The symbols are shown in the chart.

Symbol	Meaning	Example
>	Is greater than	7 > 4
<	Is less than	4 < 7
≥	Is greater than or equal to	$x \geq 3$
≤	Is less than or equal to	$-2 \leq x$

The equation $x = 5$ has one solution, 5. The inequality $x > 5$ has an infinite number of solutions: 5.001, 5.2, 6, 15, 197, and 955 are just some of the solutions. Note that 5 is not a solution—5 is not greater than 5. Because you cannot list all of the solutions, you can show them on a number line.

To show all the values that are greater than 5, but not including 5, use an open circle on 5 and shade the number line to the right.

$x > 5$

The inequality $y \leq -1$ also has an infinite number of solutions: $-1.01, -1.5, -2, -8$, and -54 are just some of the solutions. Note that -1 is also a solution, because -1 is less than *or* equal to -1. On a number line, you want to show all the values that are less than or equal to -1. Because the -1 is to be included, use a closed (filled-in) circle on -1 and shade the number line to the left.

$$y \leq -1$$

 Check It Out
Draw the number line showing the solutions to each inequality.
1. $x \geq 3$
2. $y < -1$
3. $n > -2$
4. $x \leq 0$

Solving Inequalities

Remember that addition and subtraction are opposite operations, as are multiplication and division. When you solve the inequality $x + 4 > 7$, the x has to be by itself on one side of the inequality symbol. You can use an opposite operation to achieve this. Because the x has 4 added to it, the opposite operation is to subtract 4. Do this to both sides of the inequality.

$$x + 4 > 7$$

$$x + 4 - 4 > 7 - 4$$

$$x > 3$$

You can use opposite operations to solve other inequalities.

Inequality	Opposite Operation	Applied to Both Sides	Result
$n - 6 \leq 4$	Add 6	$n - 6 + 6 \leq 4 + 6$	$n \leq 10$
$5n \geq 15$	Divide by 5	$\dfrac{5n}{5} \geq \dfrac{15}{5}$	$n \geq 3$
$\dfrac{n}{3} < 8$	Multiply by 3	$\dfrac{n}{3} \cdot 3 < 8 \cdot 3$	$n < 24$
$n + 8 > 2$	Subtract 8	$n + 8 - 8 > 2 - 8$	$n > -6$

Check It Out

Solve each inequality.

5. $x + 3 > 8$

6. $3n \leq 12$

7. $y - 7 < 2$

8. $\frac{x}{4} \geq 2$

Orphaned Whale Rescued

An orphaned baby gray whale arrived at an aquarium in California. Rescue workers named her J.J. She was three days old, weighed 1,600 lb, and was desperately ill. Within 27 days, on a diet of whale milk formula, she weighed 2,378 lb. J.J. was gaining 20 to 30 lb a day! An adult gray whale weighs approximately 35 tons, but J.J. could be released once she had a solid layer of blubber—when she weighed about 9,000 lb.

Write an equation that shows J.J. is 2,378 lb now, and needs to gain 25 lb per day for some number of days until she weighs 9,000 lb. See Hot Solutions for answer.

6·5 EXERCISES

Choose a symbol from $<, >, \leq,$ and $\geq$ for each blank.

1. 3 ___ 7
2. -8 ___ 4
3. 6 ___ 6
4. -2 ___ -7

Draw the number line showing the solutions to each inequality.

5. $x < 3$
6. $y \geq -1$
7. $n > 2$
8. $x \leq -4$

Solve each inequality.

9. $x - 4 < 5$
10. $2y \geq 8$
11. $n + 7 > 3$
12. $\frac{a}{3} \leq 6$
13. $3x \geq 15$
14. $x - 7 < 10$
15. $\frac{n}{8} \leq 1$
16. $5 + y > 6$

17. Which inequality has its solutions represented by

A. $x < 3$
B. $x \leq 3$
C. $x > 3$
D. $x \geq 3$

18. Which inequality would you solve by adding 5 to both sides?

A. $x + 5 > 9$
B. $5x < 10$
C. $\frac{x}{5} > 2$
D. $x - 5 < 4$

19. Which of the following statements is false?

A. $-7 \leq 2$
B. $0 \leq -4$
C. $8 \geq -8$
D. $5 \geq 5$

20. Which inequality would you graph by using an open circle and shading the number line to the right?

A. $x < 6$
B. $x \leq 2$
C. $x > -1$
D. $x \geq 2$

6·6 Graphing on the Coordinate Plane

Axes and Quadrants

When you cross a **horizontal** (left to right) number line with a **vertical** (up and down) number line, the result is a two-dimensional coordinate plane.

The number lines are called **axes**. The horizontal number line is the **x-axis**, and the vertical number line is the **y-axis**. The plane is divided into four regions, called **quadrants**. Each quadrant is named by a Roman numeral, as shown in the diagram.

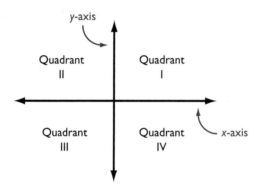

Check It Out

Fill in the blank.

1. The horizontal number line is called the ___.
2. The upper left region of the coordinate plane is called ___.
3. The lower right region of the coordinate plane is called ___.
4. The vertical number line is called the ___.

Writing an Ordered Pair

Any location on the coordinate plane can be represented by a **point.** The location of any point is given in relation to where the two axes intersect, called the **origin.**

Two numbers are required to identify the location of a point. The *x*-coordinate tells how far to the left or right of the origin the point lies. The *y*-coordinate tells how far up or down from the origin the point lies. Together, the *x*-coordinate and *y*-coordinate form an **ordered pair, (*x*, *y*).**

Since point *R* is 4 units to the right of the origin and 1 unit up, its ordered pair is (4, 1). Point *S* is 4 units to the left of the origin and 2 units down, so its ordered pair is (−4, −2).

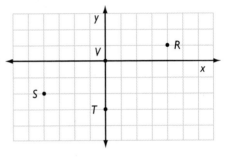

Point *T* is 0 units from the origin and 3 units down, so its ordered pair is (0, −3). Point *V* is the origin and its ordered pair is (0, 0).

 Check It Out

Give the ordered pair for each point.

5. *M*
6. *N*
7. *P*
8. *Q*

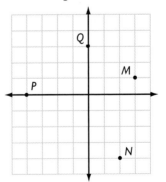

Locating Points on the Coordinate Plane

To locate point $A(3, -4)$ from the origin, move 3 units to the right and 4 units down. Point A lies in Quadrant IV. To locate point $B(-1, 4)$ from the origin, move 1 unit to the left and 4 units up. Point B lies in Quadrant II. Point $C(5, 0)$ is, from the origin, 5 units to the right and 0 units up or down. Point C lies on the x-axis. Point $D(0, -2)$ is, from the origin, 0 units to the left or right and 2 units down. Point D lies on the y-axis.

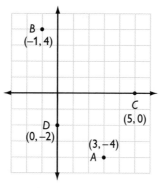

Check It Out

Locate each point on the coordinate plane and tell where it lies.

9. $H(4, -1)$
10. $J(-1, 4)$
11. $K(-2, -1)$
12. $L(0, 2)$

The Graph of an Equation with Two Variables

Consider the equation $y = 2x - 1$. Notice that it has two variables, x and y. Point $(3, 5)$ is a **solution** of this equation. If you substitute 3 for x and 5 for y (in the ordered pair, 3 is the x-coordinate and 5 is the y-coordinate), the true statement $5 = 5$ is obtained. Point $(2, 4)$ is not a solution of the equation. Substituting 2 for x and 4 for y results in the false statement $4 = 3$.

You can generate ordered pairs that are solutions.

Choose a value for x.	Substitute the value into the equation $y = 2x - 1$.	Solve for y.	Ordered Pair (x, y)
0	$y = 2(0) - 1$	-1	$(0, -1)$
1	$y = 2(1) - 1$	1	$(1, 1)$
5	$y = 2(5) - 1$	9	$(5, 9)$
-1	$y = 2(-1) - 1$	-3	$(-1, -3)$

If you locate the points on a coordinate plane, you will notice that they all lie along a straight line.

The coordinates of any point on the line will result in a true statement if they are substituted into the equation.

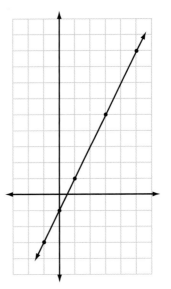

GRAPHING THE EQUATION OF A LINE

Graph the equation $y = \frac{1}{3}x - 2$.

• Choose five values for x.

Since the value of x is to be multiplied by $\frac{1}{3}$, choose values that are multiples of 3, such as $-3, 0, 3, 6,$ and 9.

• Calculate the corresponding values for y.

When $x = -3$, $y = \frac{1}{3}(-3) - 2 = -3$.

When $x = 0$, $y = \frac{1}{3}(0) - 2 = -2$.

When $x = 3$, $y = \frac{1}{3}(3) - 2 = -1$.

When $x = 6$, $y = \frac{1}{3}(6) - 2 = 0$.

When $x = 9$, $y = \frac{1}{3}(9) - 2 = 1$.

• Write the five solutions as ordered pairs (x, y).

$(-3, -3), (0, -2), (3, -1), (6, 0),$ and $(9, 1)$

• Locate the points on a coordinate plane and draw the line.

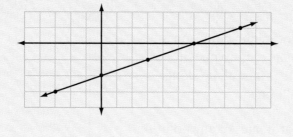

6·6 GRAPHING

◂Check It Out

Find five solutions of each equation. Graph each line.

13. $y = 3x - 2$

14. $y = 2x + 1$

15. $y = \frac{1}{2}x - 1$

16. $y = -2x + 3$

Horizontal and Vertical Lines

Choose several points that lie on a horizontal line.

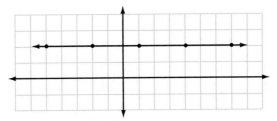

Notice that any point that lies on the line has a y-coordinate of 2. The equation of this line is $y = 2$.

Choose several points that lie on a vertical line.

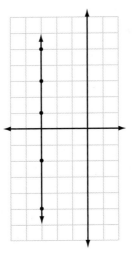

Notice that any point that lies on the line has an x-coordinate of -3. The equation of this line is $x = -3$.

 Check it Out
Graph each line.
17. $x = 3$ 18. $y = -2$
19. $x = -1$ 20. $y = 4$

Where's Your Antipode?

Imaginary lines of latitude and longitude cover the globe with a coordinate grid that is used to locate any place on the earth's surface. Latitude lines circle the globe in an east-west direction. They are drawn parallel to the equator starting at 0° and running to 90°N or 90°S at the poles. Longitude lines circle the globe in a north-south direction, meeting at both poles. These lines run from 0° at the prime meridian to 180°E or 180°W, halfway around the earth. By using latitude and longitude as coordinates, you can pinpoint any location.

Antipodes (an-ti-pə-dēz) are places opposite each other on the globe. To find your antipode (an-tə-pōd), first find the latitude of your location and change the direction. For example, if your latitude is 56°N, the latitude of your antipode would be 56°S. Next find your longitude, subtract it from 180, and change the direction. For example, if your longitude is 120°E, the longitude of your antipode would be 180 – 120 = 60°W.

Use a map to find the coordinates of your city. Find the coordinates of its antipode. Then locate your city and its antipode on a globe or map of the world.

6·6 EXERCISES

Fill in the blank.
1. The vertical number line is called the _____.
2. The lower left region of the coordinate plane is called
   _____.
3. The upper right region of the coordinate plane is called
   _____.

Give the ordered pair for each point.
4. A
5. B
6. C
7. D
8. E
9. F

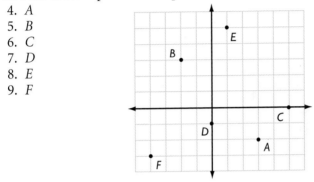

Locate each point on the coordinate plane and tell where it lies.
10. $H(-3, -4)$
11. $J(0, 2)$
12. $K(4, -1)$
13. $L(-1, 0)$
14. $M(-3, 5)$
15. $N(3, 2)$

Find five solutions for each equation. Graph each line.
16. $y = 2x - 3$
17. $y = -3x + 4$
18. $y = \frac{1}{2}x + 1$

Graph each line.
19. $x = 3$
20. $y = -4$

What have you learned?

You can use the problems and list of words below to see what you have learned in this chapter. To find out more about a particular problem or word, refer to the boldfaced topic number (for example, **6•2**).

Problem Set

Write an expression for each phrase.   **6•1**
1. a number decreased by 5
2. the product of 8 and some number
3. twice the sum of a number and 6
4. 4 less than the quotient of a number and 7

Write an equation for each sentence.   **6•1**
5. If 8 is subtracted from twice a number, the result is 4 more than the number.
6. 5 times the sum of a number and 3 is 6 less than twice the number.

Factor out the greatest common factor in each expression.   **6•2**
7. $7x + 21$ 8. $12n - 30$
9. $10a - 40$

Simplify each expression.   **6•2**
10. $10x + 3 - 6x$ 11. $8a + 3b - 5a - b$
12. $6(2n - 1) - (n + 1)$

13. Find the distance traveled by a bicyclist who rides at 18 mi/hr for $2\frac{1}{2}$ hr. Use the formula $d = rt$.   **6•3**

Use a proportion to solve items 14–15.   **6•4**
14. In a class, the ratio of boys to girls is $\frac{3}{4}$. If there are 24 girls in the class, how many boys are there?
15. A map is drawn using a scale of 120 km to 1 cm. The distance between two cities is 900 km. How far apart are the two cities on the map?

Solve each inequality. Graph the solution.   **6•5**

16. $x - 4 \leq 2$ 17. $4x > 12$

18. $n + 8 \geq 6$ 19. $\frac{n}{4} < -1$

Locate each point on the coordinate plane and tell where it lies.   **6•6**

20. $A(-3, 2)$ 21. $B(4, 0)$

22. $C(-2, -4)$ 23. $D(3, 4)$

24. $E(0, -2)$ 25. $F(-4, -1)$

Find five solutions of each equation. Graph each line.   **6•6**

26. $y = x - 3$

27. $y = -2x + 5$

28. $y = \frac{-1}{2}x + 1$

Graph each line.   **6•6**

29. $x = 1$

30. $y = -4$

---

hot **words**

WRITE DEFINITIONS FOR THE FOLLOWING WORDS.

associative
  property   **6•2**
axes   **6•6**
commutative
  property   **6•2**
cross product   **6•4**
difference   **6•1**
distributive
  property   **6•2**
equation   **6•1**
equivalent   **6•1**

equivalent
  expression   **6•2**
expression   **6•1**
formula   **6•3**
horizontal   **6•6**
inequality   **6•5**
like terms   **6•2**
order of
  operations   **6•3**
ordered pair   **6•6**
origin   **6•6**
perimeter   **6•3**
point   **6•6**
product   **6•1**

proportion   **6•4**
quadrant   **6•6**
quotient   **6•1**
rate   **6•4**
ratio   **6•4**
solution   **6•6**
sum   **6•1**
term   **6•1**
variable   **6•1**
vertical   **6•6**
x-axis   **6•6**
y-axis   **6•6**

WHAT HAVE YOU LEARNED?

Geometry

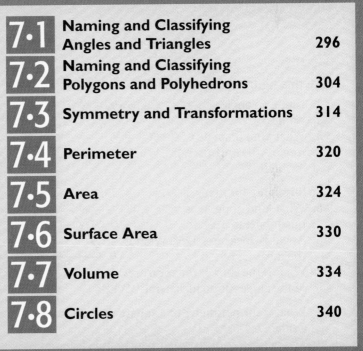

Problem Set

Use this figure for items 1–4. **7•1**

1. Name an angle.
2. Name a ray.
3. What kind of angle is $\angle ABC$?
4. Angle ABD measures 45°. What is the measure of $\angle DBC$?

Use this figure for items 5–7. **7•2**

5. What kind of figure is quadrilateral $WXYZ$?
6. Angle W measures 125°. What is the measure of $\angle Y$?
7. What is the sum of the measures of the angles of quadrilateral $WXYZ$?

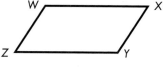

8. What is the perimeter of a square that measures 4 ft on a side? **7•4**
9. The perimeter of a regular pentagon is 150 mm. What is the length of each side? **7•4**
10. Find the area of a rectangle with a length of 12 cm and a width of 7 cm. **7•5**
11. Find the area of a right triangle whose sides measure 9 in., 12 in., and 15 in. **7•5**

12. Each of the two bases and four faces of a cube has an area of 36 mm². What is the surface area of the cube? **7•6**
13. Each face of a pyramid is a triangle with base 8 in. and height 12 in. If the area of its square base is 64 in.², what is the surface area of the pyramid? **7•6**

For items 14–16 write the letter of the polyhedron, *A*, *B*, or *C*, to match it to its name. **7•2**

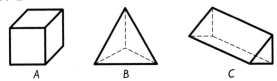

14. triangular pyramid
15. cube
16. triangular prism

17. Find the volume of a rectangular prism whose base measures 6 cm by 2 cm and whose height is 5 cm. **7•7**
18. The base of a cylinder has an area of 19.6 mm². If its height is 4 mm, what is the volume of the cylinder? **7•7**

Use circle *T* for items 19 and 20. **7•8**
19. What is the measure of $\overset{\frown}{SU}$?
20. What is the area of circle *T* rounded to the nearest tenth of an inch?

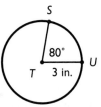

CHAPTER 7

hot **words**

angle **7•1**
arc **7•8**
base **7•5**
circle **7•6**
circumference **7•6**
congruent **7•1**
cube **7•2**
cylinder **7•6**
degree **7•1**
diagonal **7•2**
diameter **7•8**
face **7•2**
hexagon **7•2**
legs of a triangle **7•5**
line **7•1**

opposite angle **7•2**
opposite side **7•3**
parallel **7•2**
parallelogram **7•2**
pentagon **7•2**
perimeter **7•4**
perpendicular **7•5**
pi **7•8**
point **7•1**
polygon **7•1**
polyhedron **7•2**
prism **7•2**
pyramid **7•2**
Pythagorean Theorem **7•4**
quadrilateral **7•2**
radius **7•8**
ray **7•1**
rectangular prism **7•2**

reflection **7•3**
regular shape **7•2**
rhombus **7•2**
right angle **7•1**
right triangle **7•4**
rotation **7•3**
segment **7•8**
surface **7•5**
symmetry **7•3**
tetrahedron **7•2**
transformation **7•3**
translation **7•3**
trapezoid **7•2**
triangular prism **7•6**
vertex **7•1**
volume **7•7**

7·1 Naming and Classifying Angles and Triangles

Points, Lines, and Rays

In the world of math, it is sometimes necessary to refer to a specific **point** in space. Simply draw a small dot with a pencil tip to represent a point. A point has no size; its only function is to show position.

Every point needs a name, so we name a point by using a single capital letter.

· A

Point A

If you draw two points on a sheet of paper, a **line** can be used to connect them. Imagine this line as being perfectly straight and continuing without end in opposite directions. It has no thickness.

Lines need names just like points do, so that we can refer to them easily. To name a line, pick any two points on the line.

$$\overset{\longleftrightarrow}{\underset{\substack{A \qquad B}}{\bullet \qquad \bullet}}$$

Line AB, or $\overleftrightarrow{AB}$

Since the length of any line is infinite, we sometimes use parts of a line. A **ray** is part of a line that extends without end in one direction. In $\overrightarrow{AB}$, which is read as "ray AB," A is the endpoint. The second point that is used to name the ray can be any point other than the endpoint. You could also name this ray AC.

Ray AB, or $\overrightarrow{AC}$

Check It Out

Use symbols to name each line or ray in two ways.

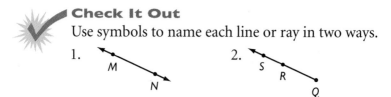

1.   M   N

2.   S   R   Q

Naming Angles

Imagine two different rays with the same endpoint. Together they form what is called an **angle.** The point they have in common is called the **vertex** of the angle. The rays form the sides of the angle.

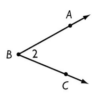

The angle above is made up of $\overrightarrow{BA}$ and $\overrightarrow{BC}$. B is the common endpoint of the two rays. Point B is the vertex of the angle. Instead of writing the word *angle,* you can use the symbol for an angle, which is $\angle$.

There are several ways to name an angle. You can name it using the three letters of the points that make up the two rays with the vertex as the middle letter ($\angle ABC$ or $\angle CBA$). You can also use just the letter of the vertex to name the angle ($\angle B$). Sometimes you might want to name an angle with a number ($\angle 2$).

When more than one angle is formed at a vertex, you use three letters to name each of the angles. Since G is the vertex of three different angles, each angle needs three letters to name it: $\angle DGF$ or $\angle FGD$, $\angle DGE$ or $\angle EGD$, $\angle EGF$ or $\angle FGE$.

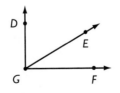

7·1 ANGLES AND TRIANGLES

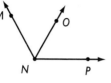 **Check It Out**

Find three angles in the figure below.

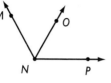

3. Name the vertex of the angles.
4. Use symbols to name each angle.

Measuring Angles

You measure an angle in **degrees,** using a *protractor* (p. 389).
The number of degrees in an angle will be greater than 0 and
less than or equal to 180.

MEASURING WITH A PROTRACTOR

To measure an angle,

- Place the center point of the protractor over the vertex
  of the angle. Line up the 0° line on the protractor with
  one side of the angle.

- Find the point where the other side of the angle crosses
  the protractor. Read the number of degrees on the scale
  at that point.

Measure ∠*ABC* and ∠*MNO*.

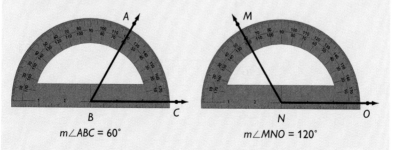

m∠ABC = 60°   m∠MNO = 120°

Check It Out
Use a protractor to measure each angle.
5. ∠*EFG*
6. ∠*PFR*
7. ∠*GFQ*

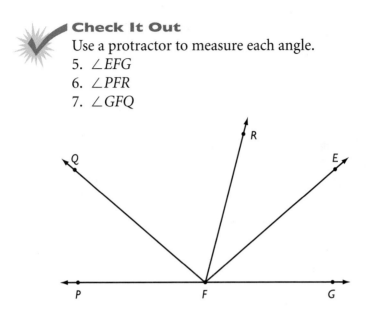

Classifying Angles

You can classify angles by their measures.

Acute angle
measures less than 90°

Right angle
measures 90°

Obtuse angle
measures greater than 90°
and less than 180°

Straight angle
measures 180°

Reflex angle
measures greater than 180°

Angles that share a side are called *adjacent angles.* You can add measures if the angles are adjacent.

$m\angle APB = 55°$
$m\angle BPC = 35°$
$m\angle APC = 55° + 35° = 90°$

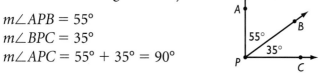

Since the sum is 90°, you know that $\angle APC$ is a **right angle.**

Check It Out

Use a protractor to measure the angle, then classify it.

8. $\angle KST$
9. $\angle RST$
10. $\angle FST$

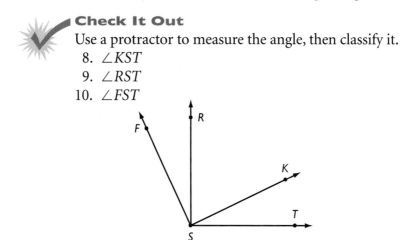

Triangles

Triangles are **polygons** that have three sides, three vertices, and three angles.

You name a triangle using the three vertices in any order. $\triangle ABC$ is read "triangle *ABC.*"

Classifying Triangles

Like angles, triangles are classified by their angle measures. They are also classified by the number of **congruent** sides, which are sides with equal length.

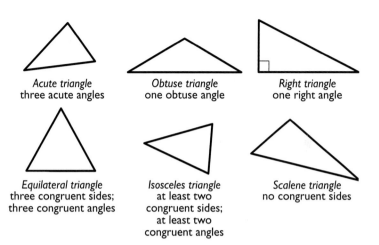

Acute triangle three acute angles	*Obtuse triangle* one obtuse angle	*Right triangle* one right angle
Equilateral triangle three congruent sides; three congruent angles	*Isosceles triangle* at least two congruent sides; at least two congruent angles	*Scalene triangle* no congruent sides

The **sum** of the measures of the three angles in a triangle is always 180°.

In △ABC, $m\angle A = 60°$, $m\angle B = 75°$, and $m\angle C = 45°$.

$60° + 75° + 45° = 180°$

So the sum of the angles of △ABC is 180°.

FINDING THE MEASURE OF THE UNKNOWN ANGLE IN A TRIANGLE

In △RST, ∠R measures 100° and ∠S measures 35°. Find the measure of ∠T.

$100° + 35° = 135°$ • Add the two angles you know.

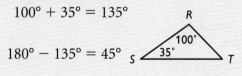

$180° - 135° = 45°$ • Subtract the sum from 180°.

$m\angle T = 45°$ • The difference is the measure of the third angle.

Check It Out

Find the measure of the third angle of each triangle.

11.
12.

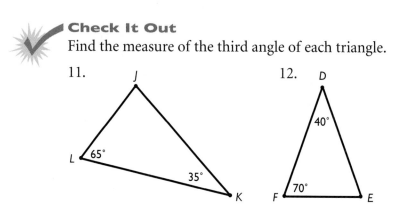

The Triangle Inequality

The length of the third side of a triangle is always less than the sum of the other two sides and greater than their difference. So $(a + b) > c > (a - b)$.

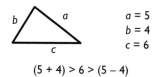

$a = 5$
$b = 4$
$c = 6$

$(5 + 4) > 6 > (5 - 4)$

Check It Out

For each of the following items, select one measurement which can be the lengths of the three sides of a triangle.

13. A. 2 in., 4 in., 8 in.
    B. 7 cm, 9 cm, 16 cm
    C. 3 m, 5 m, 4 m
14. A. 6 ft, 8 ft, 10 ft
    B. 11 m, 8 m, 2 m
    C. 14 cm, 7 cm, 7 cm

7·1 EXERCISES

Use the symbols to name each line in two ways.

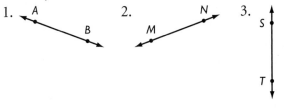

1.
2.
3.

Name each ray in two ways.

4.
5.
6.

Find three angles in this figure.
  7. Name the vertex of each angle.
  8. Use symbols to name each angle.
  9. Name the acute angle.
 10. Name the right angle.
 11. Name the obtuse angle.

12. What is the measure of ∠Q?
13. What kind of triangle is △QRS: acute, right, or obtuse?
14. Two sides of a triangle measure 7 in. and 5 in. The measure of the third side must be less than what distance?
15. Classify ∠Q.

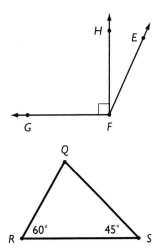

7·2 Naming and Classifying Polygons and Polyhedrons

Quadrilaterals

You may have noticed that there is a wide variety of four-sided figures, or **quadrilaterals,** to work with in geometry. All quadrilaterals have four sides and four angles. The sum of the angles of a quadrilateral is 360°. There are also many different types of quadrilaterals, which are classified by their sides and angles.

To name a quadrilateral, list the four vertices, either clockwise or counterclockwise. The quadrilateral to the right is called *FISH*

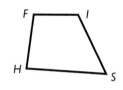

Angles of a Quadrilateral

Remember, the sum of the angles of a quadrilateral is 360°. If you know the measures of three angles in a quadrilateral, you can find the measure of the fourth angle.

FINDING THE MEASURE OF THE UNKNOWN ANGLE IN A QUADRILATERAL

Find the measure of $\angle S$ in quadrilateral *STUV.*

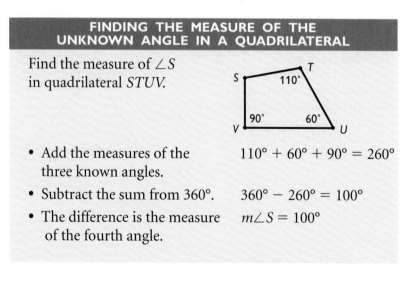

- Add the measures of the three known angles. $110° + 60° + 90° = 260°$

- Subtract the sum from 360°. $360° - 260° = 100°$

- The difference is the measure of the fourth angle. $m\angle S = 100°$

Check It Out

1. Name the quadrilateral in two ways.
2. What is the sum of the measures of ∠*M*, ∠*N*, and ∠*O*?
3. Find *m*∠*L*.

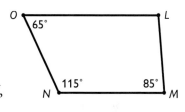

Types of Quadrilaterals

A rectangle is a quadrilateral with four right angles. *WXYZ* is a rectangle. Its length is 5 cm and its width is 3 cm.

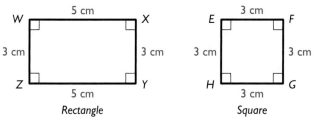

Opposite sides of a rectangle are equal in length. If all four sides of the rectangle are equal, the rectangle is called a *square.* A square is a **regular shape** because all of its sides are of equal length and all of the interior angles are *congruent.* Some rectangles may be squares, but *all* squares are rectangles. So *EFGH* is both a square and a rectangle.

A **parallelogram** is a quadrilateral with opposite sides that are **parallel.** In a parallelogram, opposite sides are equal, and **opposite angles** are equal. *ABCD* is a parallelogram. *HIJK* is both a parallelogram and a rectangle.

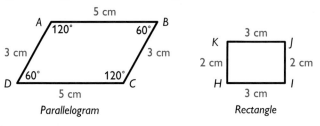

Some parallelograms may be rectangles, but *all* rectangles are parallelograms. Therefore squares are also parallelograms. If all four sides of a parallelogram are the same length, the parallelogram is called a **rhombus.** *HIJK* is a rhombus.

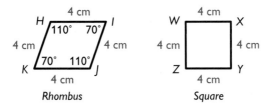

Rhombus

Square

Every square is a rhombus, although not every rhombus is a square, because a square also has equal angles.

In a **trapezoid,** two sides are parallel and two are not. A trapezoid is a quadrilateral, but it is not a parallelogram. *PARK* is a trapezoid.

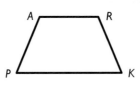

Check It Out

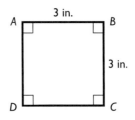

4. What kind of quadrilateral is *ABCD?*
5. What is *m∠A? m∠B? m∠C? m∠D?*
6. What is the measure of side *AD?* of side *CD?*

Polygons

A polygon is a closed figure that has three or more sides. Each side is a line segment, and the sides meet only at the endpoints, or vertices.

This figure is a polygon.   These figures are not polygons.

A rectangle, a square, a parallelogram, a rhombus, a trapezoid, and a triangle are all polygons.

There are some aspects of polygons that are always true. For example, a polygon of *n* sides has *n* angles and *n* vertices. A polygon with three sides has three angles and three vertices. A polygon with eight sides has eight angles and eight vertices, and so on.

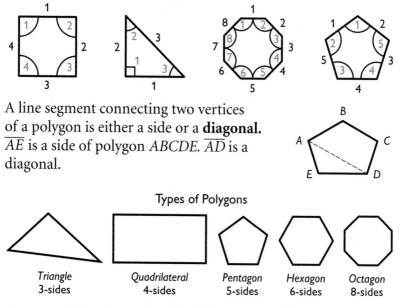

A line segment connecting two vertices of a polygon is either a side or a **diagonal.** $\overline{AE}$ is a side of polygon *ABCDE*. $\overline{AD}$ is a diagonal.

Types of Polygons

Triangle 3-sides	Quadrilateral 4-sides	Pentagon 5-sides	Hexagon 6-sides	Octagon 8-sides

A seven-sided polygon is called a **heptagon,** a nine-sided polygon is called a **nonagon,** and a ten-sided polygon is called a *decagon.*

7-2 POLYGONS AND POLYHEDRONS

Check It Out

Name each polygon.

7. 8. 9. 10.

Oh, Obelisk!

Ancient Egyptian obelisks were carved horizontally out of the rock quarry. Exactly how the Egyptians lifted the obelisks into vertical position is a mystery. But clues suggest that the Egyptians slid them down a dirt ramp, inched them higher with levers, and finally pulled them upright with ropes.

A crew from a television station attempted to move a 43-ft-long block of granite using this method. They tilted the 40-ton obelisk down a ramp at a 33° angle. With levers, they inched the obelisk up to about a 40° angle. Then 200 people tried to haul it with ropes to a standing position. They couldn't budge it. Finally, out of time and money, they abandoned the attempt.

How many additional degrees did the crew need to raise the obelisk before it would have stood upright? See Hot Solutions for answer.

Angles of a Polygon

You know that the sum of the angles of a triangle is 180° and that the sum of the angles of a quadrilateral is 360°. The sum of the angles of *any* polygon totals at least 180° (triangle). Each additional side adds 180° to the measure of the first three angles. To see why, look at a **pentagon.**

Drawing diagonals $\overline{EB}$ and $\overline{EC}$ shows that the sum of the angles of a pentagon is the sum of the angles in three triangles.

$$3 \times 180° = 540°$$

So the sum of the angles of a pentagon is 540°.

You can use the formula $(n - 2) \times 180°$ to find the sum of the angles of a polygon. Just let n equal the number of sides of a polygon. The answer you get is the sum of the measures of all the angles of the polygon.

FINDING THE SUM OF THE ANGLES OF A POLYGON

$(n - 2) \times 180° =$ sum of polygon with n sides

Find the sum of the angles of an octagon.

Think: An octagon has 8 sides. Subtract 2. Then multiply the difference by 180.

• Use the formula: $(8 - 2) \times 180° = 6 \times 180° = 1,080°$

So the sum of the angles of an octagon is 1,080°.

As you know, a **regular polygon** has equal sides and equal angles. You can use what you know about finding the sum of the angles of a polygon to find the measure of each angle of a regular polygon.

Find the measure of each angle in a regular **hexagon.**

Begin by using the formula $(n - 2) \times 180°$. A hexagon has 6 sides, and so you should substitute 6 for n.

$$(6 - 2) \times 180° = 4 \times 180° = 720°$$

Then divide the sum of the angles by the number of angles. Since a hexagon has 6 angles, divide by 6.

$$720° \div 6 = 120°$$

The answer tells you that each angle of a regular hexagon measures 120°.

Check It Out

11. What is the sum of the angles of a heptagon?
12. What is the measure of each angle of a regular pentagon?

Polyhedrons

Solid shapes can be curved, like these.

Sphere   Cylinder   Cone

Some solid shapes have flat surfaces. Each of the figures below is a **polyhedron.**

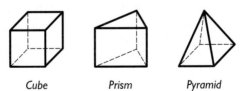

Cube   Prism   Pyramid

A polyhedron is any solid whose surface is made up of polygons. Triangles, quadrilaterals, and pentagons make up the **faces** of the common polyhedrons below.

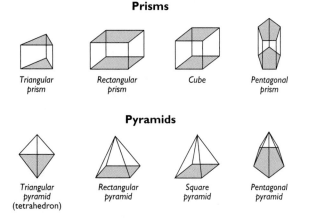

Prisms

Triangular prism	Rectangular prism	Cube	Pentagonal prism

Pyramids

Triangular pyramid (tetrahedron)	Rectangular pyramid	Square pyramid	Pentagonal pyramid

A **prism** has two bases, or "end" faces. The bases of a prism are the same size and shape and are parallel to each other. Its other faces are parallelograms. The bases of each prism are shaded in the chart. When all six faces of a **rectangular prism** are square, the figure is a **cube.**

A **pyramid** is a structure that has one base in the shape of a polygon. It has triangular faces that meet a point called the *apex.* The base of each pyramid is shaded in the chart.
A triangular pyramid is a **tetrahedron.** A tetrahedron has four faces. Each face is triangular. A triangular prism, however, is *not* a tetrahedron.

> ✔ **Check It Out**
> Identify each polyhedron.
> 13.   14.

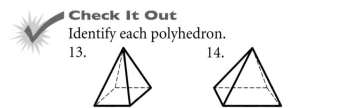

7·2 EXERCISES

1. Find the measure of ∠J.
2. Give two other names for quadrilateral *IJKL*.

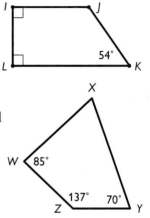

3. Find the measure of ∠X.
4. Give two other names for quadrilateral *WXYZ*.

Write the letter of the word in the word box that matches each description.

5. It is a solid figure whose surface is made up of polygons.
6. Its base is a triangle. Its faces are triangles.
7. Its base is a square. Its faces are triangles.
8. It is a quadrilateral with opposite sides parallel and opposite angles equal.
9. It is a closed figure that has three or more sides.
10. It is a rectangular prism with six square faces.
11. It has two triangular bases and faces that are parallelograms.

WORD BOX
A. Rhombus
B. Polyhedron
C. Triangular prism
D. Square pyramid
E. Cube
F. Polygon
G. Triangular pyramid

12. Give two other names for quadrilateral *EFGH*.
13. Find the measure of ∠G.

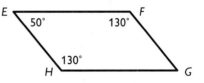

Identify each polygon.

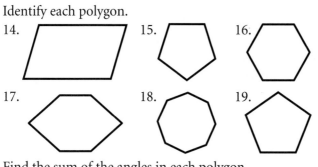

14.   15.   16.

17.   18.   19.

Find the sum of the angles in each polygon.

20. square   21. pentagon   22. octagon

Two angles of $\triangle XYZ$ measure 23° and 67°.

23. What is the measure of the third angle?
24. What kind of triangle is $\triangle XYZ$?

Identify each polyhedron.

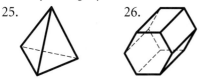

25.   26.

27. Name the figure that has two identical bases that are parallel pentagons.
28. Name the figure that has a square base and faces that are triangles.

Identify the shape of these real-world polyhedrons.

29.   30.

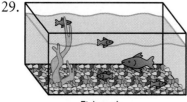

Fish tank

Nut

7·3 Symmetry and Transformations

Whenever you move a shape that is in a plane, you are performing a **transformation.**

Reflections

A **reflection** (or **flip**) is one kind of transformation. When you hear the word "reflection," you may think of a mirror. The mirror image, or reverse image, of a point or shape is called a *reflection.*

The reflection of a point is another point on the other side of a line of **symmetry.** Both the point and its reflection are the same distance from the line.

P' reflects point P on the other side of line l. P' is read "P-prime." P' is called the *image* of P.

Any point, line, or polygon can be reflected. Quadrilateral *DEFG* is reflected on the other side of line m. The image of *DEFG* is $D'E'F'G'$.

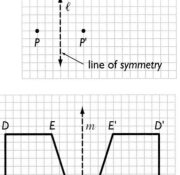

To find an image of a shape, pick several key points in the shape. For a polygon, use the vertices. For each point, measure the distance to the line of symmetry. The image of each point will be the same distance from the line of symmetry on the opposite side.

In the quadrilateral reflection on the opposite page, point *D* is 10 units from the line of symmetry, and point *D'* is also 10 units from the line on the **opposite side.** You can measure the distance from the line for each point, and the corresponding image point will be the same distance.

Check It Out

1. Draw figure *ABCD* and line *e* on grid paper. Then draw and label the image of the figure.

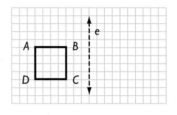

Fish Farming

Fish don't grow on trees.

Fish farming, one of the fastest growing food industries, now supplies about 20 percent of the world's fish and shellfish.

A Japanese oyster farmer builds big floating bamboo rafts in the ocean, then hangs clean shells from them by ropes. Oyster larva attach to the shells and grow in thick masses. The rafts are supported with barrels so they don't sink to the bottom where the oysters' natural predators, starfish, can get them.

An oyster farmer might have 100 rafts, each about 10 m by 15 m. What is the total area of the rafts? See Hot Solutions for answer.

Reflection Symmetry

You have seen that a line of symmetry is used to show the reflection symmetry of a point, a line, or a shape. A line of symmetry can also *separate* a shape into two parts, where one part is a reflection of the other. Each of these figures is symmetrical with respect to the line of symmetry.

Sometimes a figure has more than one line of symmetry. Here are more shapes that have more than one line of symmetry.

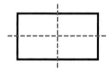

A rectangle has two lines of symmetry.

A square has four lines of symmetry.

Any line that passes through the center of a circle is a line of symmetry. So a circle has an infinite number of lines of symmetry.

 Check It Out

Some capital letters have reflection symmetry.

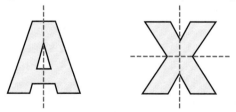

2. The letter A has one line of symmetry. Which other capital letters have exactly one line of symmetry?
3. The letter X has two lines of symmetry. Which other capital letters have two lines of symmetry?

Rotations

A **rotation** (or **turn**) is a transformation that turns a line or a shape around a fixed point. This point is called the *center of rotation*. Degrees of rotation are usually measured in the counterclockwise direction.

$\overleftrightarrow{RS}$ is rotated 90° around point R.

If you rotate a figure 360°, it comes back to where it started. Despite the rotation, its position is unchanged.
If you rotate $\overrightarrow{AB}$ 360° around point P, $\overrightarrow{AB}$ is still in the same place.

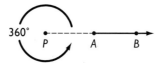

Check It Out

How many degrees has the flag been rotated around point *H* or *J*?

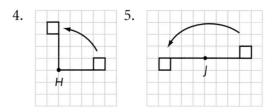

4.   5.

Translations

A **translation** (or **slide**) is another kind of transformation. When you slide a figure to a new position without turning it, you are performing a translation.

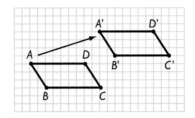

Rectangle *ABCD* moves right and up. *A'B'C'D'* is the image of *ABCD* under a translation. *A'* is 9 units to the right and 4 units up from *A*. All other points on the rectangle moved the same way.

 Check It Out

6. Which of the figures is a translation of the shaded figure?

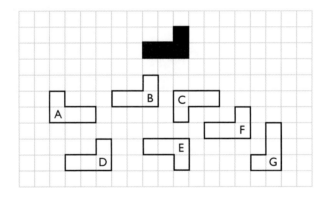

**EXERCISES**

What is the reflected image across line *s* of each of the following?

1. Point *D*   2. $\overline{DF}$ 3. △ *DEF*

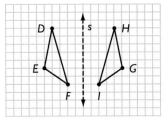

Trace each shape. Then draw all lines of symmetry.

For each transformation, tell whether the image shows a flip, a turn, or a slide.

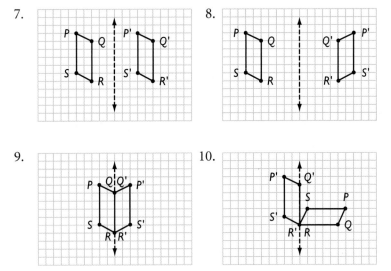

7·4 Perimeter

Perimeter of a Polygon

Ramon is planning to put a fence around his pasture. To
determine how much fencing he needs, he must calculate the
perimeter of, or *distance around*, his pasture.

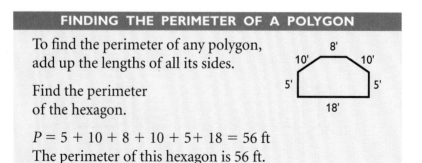

RAMON'S PASTURE

The perimeter
of any polygon
is the sum of the
lengths of the sides of
the polygon. So to find
the perimeter of his pasture, Ramon needs to measure the
length of each side and to find the sum. He finds that two of
the sides are 120 yd, one side is 180 yd, and another is 150 yd.
How much fencing will Ramon need to enclose his pasture?

$$P = 120 \text{ yd} + 150 \text{ yd} + 120 \text{ yd} + 180 \text{ yd} = 570 \text{ yd}$$

The perimeter of the pasture is 570 yd. Ramon will need
570 yd of fencing to enclose the pasture.

FINDING THE PERIMETER OF A POLYGON

To find the perimeter of any polygon,
add up the lengths of all its sides.

Find the perimeter
of the hexagon.

$P = 5 + 10 + 8 + 10 + 5 + 18 = 56$ ft
The perimeter of this hexagon is 56 ft.

Regular Polygon Perimeters

The sides of a regular polygon are all the same length. If you know the perimeter of a regular polygon, you can find the length of each side.

To find the length of each side of a regular octagon with a perimeter of 36 cm, let x = length of a side.

$$36 \text{ cm} = 8x$$
$$4.5 \text{ cm} = x$$

Each side is 4.5 cm long.

Perimeter of a Rectangle

Opposite sides of a rectangle are equal in length. So, to find the perimeter of a rectangle, you only need to know its length and width. The formula for the perimeter of a rectangle is $2l + 2w = P$. The perimeter of this rectangle is $(2 \times 7 \text{ cm}) + (2 \times 3 \text{ cm}) = 20 \text{ cm}$.

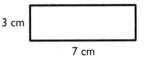
3 cm

7 cm

✔ Check It Out

Find the perimeter of each polygon.

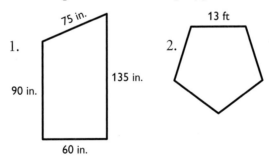

1. 75 in. / 90 in. / 135 in. / 60 in.

2. 13 ft

3. Find the perimeter of a rectangle with length 43 in. and width 15 in.
4. Find the perimeter of a square with sides that measure 6 km.

Perimeter of a Right Triangle

If you know the lengths of two sides of a **right triangle,** you can find the length of the third side by using the **Pythagorean Theorem.**

FINDING THE PERIMETER OF A RIGHT TRIANGLE

Use the Pythagorean Theorem to find the perimeter of a right triangle.

- Use the equation $c^2 = a^2 + b^2$ to find the length of the hypotenuse.

  $a = 16$ cm
  $b = 30$ cm

  $$c^2 = 16^2 + 30^2$$
  $$= 256 + 900$$
  $$= 1,156$$

- The square root of c^2 is equal to the length of the hypotenuse. So find the square root of c^2.

  $$c = 34$$

- Add the lengths of the sides. The sum is the perimeter of the triangle.

  $$16 \text{ cm} + 30 \text{ cm} + 34 \text{ cm} = 80 \text{ cm}$$

The perimeter is 80 cm.

✓ Check It Out

Use the Pythagorean Theorem to find the perimeter of each triangle.

5.

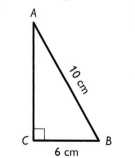

6.

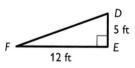

7·4 EXERCISES

Find the perimeter of each polygon.

1.

10 in.   15 in.   8 in.

2.   12 ft   10 ft   19 ft   6 ft

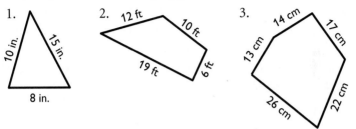

3.   14 cm   17 cm   13 cm   26 cm   22 cm

4. Find the perimeter of a square that measures 19 m on a side.
5. Find the perimeter of a regular hexagon that measures 6 cm on a side.
6. The perimeter of a regular pentagon is 140 in. Find the length of each side.
7. The perimeter of an equilateral triangle is 108 mm. What is the length of each side?

Find the perimeter of each rectangle.
8. $l = 28$ m, $w = 11$ m
9. $l = 25$ yd, $w = 16$ yd
10. $l = 43$ ft, $w = 7$ ft

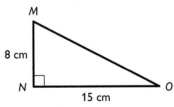

M

8 cm

N   15 cm   O

11. Find the length of $\overline{MO}$.
12. Find the perimeter of $\triangle MNO$.

Use the Pythagorean Theorem to tell whether the triangle is a right triangle.
13. $\triangle OPQ$ has sides that measure 8, 6, 10.
14. $\triangle RST$ has sides that measure 7, 10, 13.
15. $\triangle WXY$ has sides that measure 5, 6, 9.

7·5 Area

What Is Area?

Area measures the size of a surface. Your desktop is a surface with area, and so is the state of Montana. Instead of measuring with units of length, such as inches, centimeters, feet, and kilometers, you measure area in square units, such as **square inches** (in.^{2}) and **square centimeters** (cm^2).

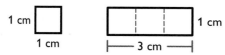

The square has an area of one square centimeter. It takes exactly three of these squares to cover the rectangle, which tells you that the area of the rectangle is three square centimeters, or 3 cm^2.

Estimating Area

When an exact answer is not needed or is hard to find, you can **estimate** the area of a surface.

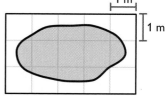

In the shaded figure (right), three squares are completely shaded, and so you know that the area is greater than 3 m^2. The rectangle around the shape covers 15 m^2, and obviously the shaded area is less than that. So you can estimate the area of the shaded figure by saying that it is greater than 3 m^2 but less than 15 m^2.

Check It Out

For each of the following items, select the one with the area of a region.
1. A. 26 ft B. 60 m C. 16 cm^2
2. A. 4 mi B. 37 yd^2 C. 80 km
3. Estimate the area of the shaded region. Each square represents 1 in.^{2}.

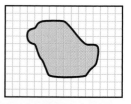

Area of a Rectangle

You can count squares to find the area of this rectangle.

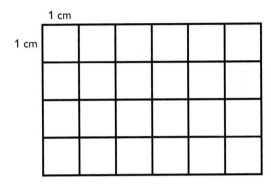

There are 24 squares and each is a square centimeter. So the area of this rectangle is 24 cm^2.

You can also use the formula for finding the area of a rectangle: $A = l \times w$. The length of the rectangle above is 6 cm and the width is 4 cm. Using the formula, you find that

$$A = 6 \text{ cm} \times 4 \text{ cm} = 24 \text{ cm}^2$$

FINDING THE AREA OF A RECTANGLE

Find the area of this rectangle.

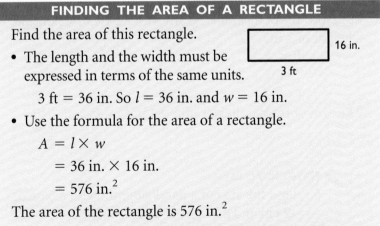

- The length and the width must be expressed in terms of the same units.

    3 ft = 36 in. So l = 36 in. and w = 16 in.
- Use the formula for the area of a rectangle.

    $A = l \times w$

    $= 36 \text{ in.} \times 16 \text{ in.}$

    $= 576 \text{ in.}^2$

The area of the rectangle is 576 in.^{2}

If the rectangle is a square, the length and the width are the same. So for a square whose sides measure s units, you can use the formula $A = s \times s$, or $A = s^2$.

Check It Out

4. Find the area of a rectangle with a length of 24 yd and a width of 17 yd.
5. Find the area of a square that measures 9 in. on a side.

Area of a Parallelogram

To find the area of a parallelogram, you multiply the **base** by the **height.**

Area = base × height
$A = b \times h$,
or $A = bh$

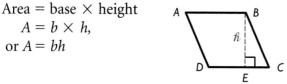

The height of a parallelogram is always **perpendicular** to the base. So in parallelogram $ABCD$, the height, h, is equal to $\overline{BE}$, not $\overline{BC}$. The base, b, is equal to $\overline{DC}$.

FINDING THE AREA OF A PARALLELOGRAM

Find the area of a parallelogram with a base of 12 in. and a height of 7 in.

$A = b \times h$
$= 12$ in. $\times 7$ in.
$= 84$ in.^{2}

The area of the parallelogram is 84 in.^{2} or 84 square in.

Check It Out

6. Find the area of a parallelogram with a base of 12 cm and a height of 15 cm.
7. Find the length of the base of a parallelogram whose area is 56 ft^2 and whose height is 7 ft.

Area of a Triangle

If you were to cut a parallelogram along a diagonal, you would have two triangles with equal bases, b, and the same height, h.

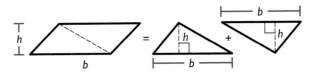

A triangle has half the area of a parallelogram with the same base and height. The area of a triangle equals $\frac{1}{2}$ the base times the height, and so the formula is $A = \frac{1}{2} \times b \times h$, or $A = \frac{1}{2}bh$.

$A = \frac{1}{2} \times b \times h$

$A = \frac{1}{2} \times 13.5 \text{ cm} \times 8.4 \text{ cm}$

$\quad = 0.5 \times 13.5 \text{ cm} \times 8.4 \text{ cm}$

$\quad = 56.7 \text{ cm}^2$

The area of the triangle is 56.7 cm^2.

FINDING THE AREA OF A TRIANGLE

Find the area of $\triangle PQR$. Note that in a right triangle the two **legs** serve as a height and a base.

$A = \frac{1}{2}bh$

$\quad = \frac{1}{2} \times 5 \text{ m} \times 3 \text{ m}$

$\quad = 0.5 \times 5 \text{ m} \times 3 \text{ m}$

$\quad = 7.5 \text{ m}^2$

The area of the triangle is 7.5 m^2.

For a review of *right triangles*, see page 322.

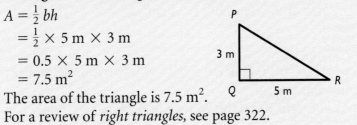

Check It Out

8. Find the area of a triangle whose base is 16 ft and whose height is 8 ft.

9. Find the area of a right triangle whose sides measure 6 cm, 8 cm, and 10 cm.

Area of a Trapezoid

A trapezoid has two bases, which are labeled b_1 and b_2. You read b_1 as "b sub-one." The area of a trapezoid is equal to the area of two triangles.

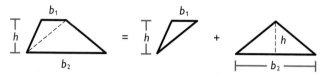

You know that the formula for the area of a triangle is $A = \frac{1}{2}bh$, and so it makes sense that the formula for finding the area of a trapezoid would be $A = \frac{1}{2}b_1h + \frac{1}{2}b_2h$ or, in simplified form, $A = \frac{1}{2}h(b_1 + b_2)$.

FINDING THE AREA OF A TRAPEZOID

Find the area of trapezoid $WXYZ$.

$A = \frac{1}{2}h(b_1 + b_2)$
$\quad = \frac{1}{2} \times 4\,(5 + 11)$
$\quad = 2 \times 16$
$\quad = 32 \text{ cm}^2$

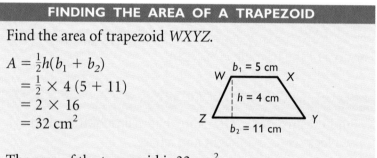

The area of the trapezoid is 32 cm^2.

Since $\frac{1}{2}h(b_1 + b_2)$ is equal to $h \times \frac{b_1 + b_2}{2}$, you can remember the formula this way:

$\quad A$ = height times the average of the bases

For a review of how to find an *average* or *mean*, see page 204.

Check It Out

10. A trapezoid has a height of 4 m. Its bases measure 5 m and 8 m. What is its area?
11. A trapezoid has a height of 8 cm. Its bases measure 7 cm and 10 cm. What is its area?

EXERCISES

1. Estimate the area of the shaded region.

2. If each of the square units covered by the region measures 2 cm², estimate the area in centimeters.

Find the area of each rectangle given length, *l*, and width, *w*.
 3. *l* = 5 ft, *w* = 4 ft
 4. *l* = 7.5 in., *w* = 6 in.

Find the area of each figure.

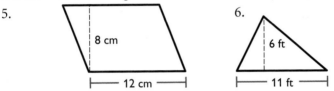

5. 8 cm, 12 cm

6. 6 ft, 11 ft

Find the area of each triangle given base, *b*, and height, *h*.
 7. *b* = 12 cm, *h* = 9 cm
 8. *b* = 8 yd, *h* = 18 yd

 9. Find the area of a trapezoid whose bases are 10 in. and 15 in. and whose height is 8 in.
10. This is a design for a new pigpen. If the farmer builds it to have these measurements, how much area inside the pen will there be for the pigs?

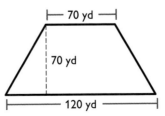

70 yd

70 yd

120 yd

7·6 Surface Area

The surface area of a solid is the total area of its exterior surfaces. You can think about surface area in terms of the parts of a solid shape that you would paint. Like area, surface area is expressed in square units. To see why, "unfold" the rectangular prism.

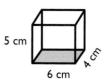

Mathematicians call this unfolded prism a *net*. A net can be folded to make a three-dimensional figure.

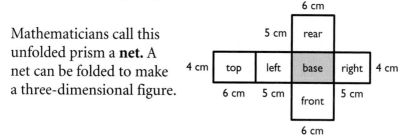

Surface Area of a Rectangular Prism

A rectangular prism has six rectangular faces. To find the surface area of a rectangular prism, find the sum of the areas of the six faces, or rectangles. Remember, opposite faces are equal.

For a review of *polyhedrons* and *prisms,* see pages 310-311.

FINDING THE SURFACE AREA OF A RECTANGULAR PRISM

Use the net to find the area of the rectangular prism above.

• Use the formula $A = lw$ to find the area of each face.

• Then add the six areas.

• Express the answer in square units.

$$
\begin{aligned}
\text{Area} &= \text{top + base} &+& \quad \text{left + right} &+& \quad \text{front + rear} \\
&= 2 \times (6 \times 4) &+& \quad 2 \times (5 \times 4) &+& \quad 2 \times (6 \times 5) \\
&= \quad 2 \times 24 &+& \quad\quad 2 \times 20 &+& \quad\quad 2 \times 30 \\
&= \quad\quad 48 &+& \quad\quad\quad 40 &+& \quad\quad\quad 60 \\
&= 148 \text{ cm}^2
\end{aligned}
$$

The surface area of the rectangular prism is 148 cm^2.

Check It Out

Find the surface area of each shape.

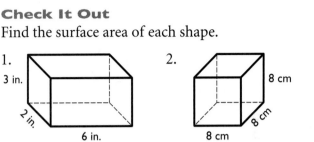

1. 3 in. / 2 in. / 6 in.

2. 8 cm / 8 cm / 8 cm

Surface Area of Other Solids

The unfolding technique can be used to find the surface area of any polyhedron. Look at the **triangular prism** and its net.

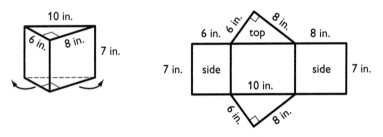

To find the surface area of this solid, use the area formulas for a rectangle ($A = lw$) and a triangle ($A = \frac{1}{2}bh$) to find the areas of the five faces and then find the sum of the areas.

Below are two pyramids and their nets. For these polyhedrons, you would again use the area formulas for a rectangle ($A = lw$) and a triangle ($A = \frac{1}{2}bh$) to find the areas of the faces and then find the sum of the areas.

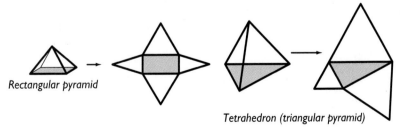

Rectangular pyramid

Tetrahedron (triangular pyramid)

The surface area of a **cylinder** is the sum of the areas of two **circles** and a rectangle.

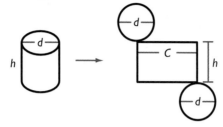

The two bases of a cylinder are equal in area. The height of the rectangle is the height of the cylinder. Its length is the **circumference** of the cylinder.

To find the surface area of a cylinder, you would:
- Use the formula for the area of a circle to find the area of each base.
  $$A = \pi r^2$$
- Find the area of the rectangle using the formula $h \times (2\pi r)$.

For a review of *circles*, see page 340.

Check It Out

3. Unfold this triangular prism and find its surface area.

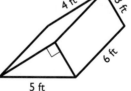

4. Find the surface area of the cylinder. Use $\pi = 3.14$.

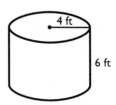

7·6 EXERCISES

Find the surface area of each shape.

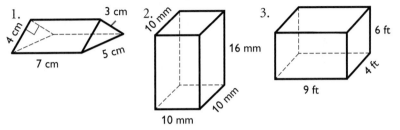

1. 3 cm 4 cm 7 cm 5 cm

2. 10 mm 16 mm 10 mm 10 mm

3. 6 ft 9 ft 4 ft

4. A rectangular prism measures 5 cm by 3 cm by 7cm. Find its surface area.

5. The surface area of a cube is 150 in². What is the length of an edge?
   A. 5 in.   B. 6 in.   C. 7 in.   D. 8 in.

Find the surface area of each cylinder. Use 3.14 for π.

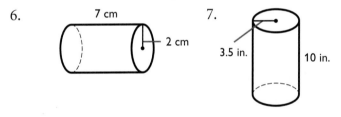

6. 7 cm 2 cm

7. 3.5 in. 10 in.

8. What is the surface area of a cube with side 3 m?

9. What is the surface area of an equilateral tetrahedron with one face equal to 25 in.²?

10. Toshi is painting a box that is 4 ft. long, 1 ft. wide, and 1 ft. high. He wants to put three coats of lacquer on it. He has a can of lacquer that will cover 200 square feet. Does he have enough lacquer for three coats?

7·7 Volume

What Is Volume?

Volume is the space inside a figure. One way to measure volume is to count the number of cubic units that would fill the space inside a figure.

The volume of this small cube is 1 cubic inch.

The number of smaller cubes that it takes to fill the space inside the larger cube is 8, and so the volume of the larger cube is 8 cubic inches.

You measure the volume of shapes in *cubic* units. For example, 1 cubic inch is written as 1 in.^{3}, and 1 cubic meter is written as 1 m^3.

✔ Check It Out

What is the volume of each shape if 1 cube = 1 in.^{3}?

1.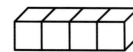

2.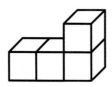

Volume of a Prism

The volume of a prism can be found by multiplying the *area* (pp. 324–329) of the base, *B*, and the height, *h*.

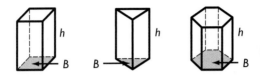

Volume = *Bh*

See *formulas*, pages 58–59.

FINDING THE VOLUME OF A PRISM

Find the volume of the rectangular prism. The base is 12 in. long and 10 in. wide; the height is 15 in.

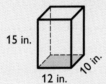

15 in.

12 in.   10 in.

base $A = 12$ in. $\times 10$ in.   • Find the area of the
    $= 120$ in.^{2} base.
$V = 120$ in.$^2 \times 15$ in.   • Multiply the base and
    $= 1,800$ in.^{3} the height.

The volume of the prism is 1,800 in.^{3}.

✔ Check It Out

Find the volume of each shape.

3.

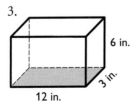

6 in.

3 in.

12 in.

4.

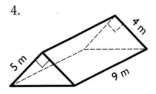

4 m

5 m

9 m

Volume of a Cylinder

You can find the volume of a cylinder the same way you found the volume of a prism, using the formula $V = Bh$. Remember that the base of a cylinder is a circle.

7 in.

2 in.

The base has a radius of 2 in. Estimate **pi** (π) at 3.14 to find the area of the base.

$$B = \pi r^2 = \pi \times 2^2 = \pi \times 4 = 12.56$$

The area of the base is about 12.56 square inches. Since you also know the height, you can use the formula $V = Bh$.

$$V = 12.56 \text{ in.}^2 \times 7 \text{ in.} = 87.92 \text{ in.}^3$$

The volume of the cylinder is 87.92 in.^{3}

Check It Out

Find the volume of each cylinder. Round to the nearest hundredth. Use 3.14 for π.

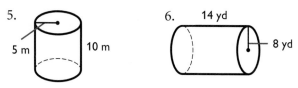

5. 5 m 10 m

6. 14 yd 8 yd

Volume of a Pyramid and a Cone

The formula for the volume of a pyramid or a cone is $V = \frac{1}{3}Bh$.

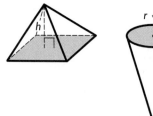

FINDING THE VOLUME OF A PYRAMID

Find the volume of the pyramid. The base is 175 cm long and 90 cm wide; the height is 200 cm.

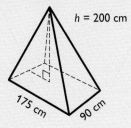

h = 200 cm

175 cm

90 cm

base $A = (175 \times 90)$ • Find the area of the base.

$= 15{,}750 \text{ cm}^2$

$V = \frac{1}{3}(15{,}750 \times 200)$ • Multiply the base by the

$= 1{,}050{,}000$ height and then by $\frac{1}{3}$.

The volume is $1{,}050{,}000 \text{ cm}^3$.

To find the volume of a cone, you follow the same procedure as above. You may use your calculator to help find the area of the base of the cone. For example, a cone has a base with a radius of 3 cm and a height of 10 cm. What is the volume of the cone to the nearest tenth?

Square the radius and multiply by π to find the area of the base. Then multiply by the height and divide by 3 to find the volume. The volume of the cone is 94.2 cm^3.

Press $\boxed{\pi}$ $\boxed{\times}$ 9 $\boxed{=}$ $\boxed{\text{28.27433}}$ $\boxed{\times}$ 10 $\boxed{\div}$ 3 $\boxed{=}$ $\boxed{\text{94.24778}}$

For other volume *formulas,* see page 58.

7·7 VOLUME

✓ Check It Out

Find the volume of the shapes below, rounded to the nearest tenth. Use 3.14 for π.

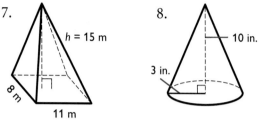

7.

$h = 15$ m

8 m

11 m

8.

10 in.

3 in.

Good Night, T. Rex

Why did the dinosaurs disappear? New evidence from the ocean floor points to a giant asteroid that collided with Earth some 65 million years ago.

The asteroid, 6 to 12 mi in diameter, hit the earth somewhere in the Gulf of Mexico. It was traveling at a speed of thousands of miles per hour.

The collision sent billions of tons of debris into the atmosphere. The debris rained down on the planet, obscuring the sun. Global temperatures plummeted. The fossil record shows that most of the species that were alive before the collision disappeared.

Assume the crater left by the asteroid had the shape of a hemisphere with a diameter of 165 mi. How many cubic miles of debris would have been flung from the crater into the air? For formula for volume of sphere, see p. 58. See Hot Solutions for answer.

7·7 EXERCISES

Use the rectangular prism to answer items 1–4.

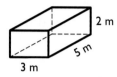

3 m

1. How many cubes would it take to make one layer in the bottom of the prism?
2. How many layers of cubes would it take to fill the prism?
3. How many cubes would it take to fill the prism?
4. Each cube has a volume of 1 cm³. What is the volume of the prism?

5. Find the volume of a rectangular prism whose base measures 3 ft by 4 ft and whose height is 6 ft.
6. The base of a cylinder has an area of 28.26 sq in. Its height is 12.5 in. What is its volume?
7. Find the volume of a cylinder that is 12 cm high and whose base has a radius of 8 cm. Use 3.14 for π.
8. Find the volume of a pyramid with a height of 9 cm and a base that is 2.5 cm wide and 7 cm long.

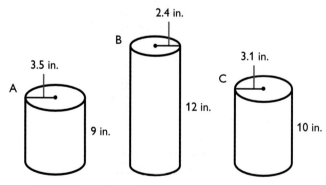

9. Which can would hold the most juice: A, B, or C?
10. Write the letters of the cans in order, from the can that holds the least to the can that holds the most.

7·8 Circles

Parts of a Circle

Of the many shapes you may encounter in geometry, circles are among the most unique. They differ from other geometric shapes in several ways. For instance, while all circles are the same shape, polygons vary in shape. Circles do not have any sides, while polygons are named and classified by the number of sides they have. The *only* thing that makes one circle different from another is size.

A circle is a set of points equidistant from a given point. That point is the **center of the circle.** A circle is named by its center point.

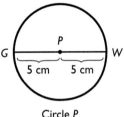

Circle P

A **radius** is a **segment** that has one endpoint at the center and the other endpoint on the circle. In circle P, $\overline{PW}$ is a *radius,* and so is $\overline{PG}$.

A **diameter** is a line segment that passes through the center of the circle and has both endpoints on the circle. $\overline{GW}$ is a diameter of circle P. Notice that the length of the diameter $\overline{GW}$ is equal to the sum of $\overline{PW}$ and $\overline{PG}$. So the diameter is twice the length of the radius. If d equals the diameter and r equals the radius, d is twice the radius, r. So the diameter of circle P is 2(5) or 10 cm.

Check It Out

1. Find the radius of a circle with diameter 12 cm.
2. Find the radius of a circle with diameter 15 ft.
3. Find the radius of a circle in which $d = y$.
4. Find the diameter of a circle with radius 10 in.
5. Find the diameter of a circle with radius 5.5 m.
6. Express the diameter of a circle whose radius is equal to x.

Circumference

The circumference of a circle is the distance around the circle. The **ratio** (p. 274) of every circle's circumference to its diameter is always the same. That ratio is a number that is close to 3.14. In other words, in every circle, the circumference is about 3.14 times the diameter. The symbol π, which is read as *pi,* is used to represent the ratio $\frac{C}{d}$.

$$\frac{C}{d} = 3.141592...$$

Circumference = pi $\times$ diameter, or $C = \pi d$

Look at the illustration below. The circumference of the circle is about the same length as three diameters. This is true for any circle.

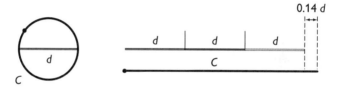

Since $d = 2r$, Circumference = 2 $\times$ pi $\times$ radius, or $C = 2\pi r$.

If you are using a calculator that has a π key, hit it, and you will get an approximation for π to several more decimal places: $\pi = 3.141592....$ For practical purposes, however, when you are finding the circumference of a circle, round π to 3.14, or simply leave the answer in terms of π.

FINDING THE CIRCUMFERENCE OF A CIRCLE

Find the circumference of a circle with radius 8 m.

- Use the formula $C = \pi d$. Remember to multiply the radius by 2 to get the diameter. Round the answer to the nearest tenth.

$$d = 8 \times 2 = 16$$
$$C = 16\pi$$

The exact circumference is 16π m.

$$C = 16 \times 3.14$$
$$= 50.24$$

To the nearest tenth, the circumference is 50.2 m.

You can find the diameter if you know the circumference. Divide both sides by π.

$$C = \pi d \qquad \frac{C}{\pi} = \frac{\pi d}{\pi} \qquad \frac{C}{\pi} = d$$

Check It Out

7. Find the circumference of a circle with a diameter of 8 mm. Give your answer in terms of π.
8. Find the circumference of a circle with a radius of 5 m. Round to the nearest tenth.
9. Find the diameter of a circle with a circumference of 44 ft to the nearest tenth.
10. Find the radius of a circle with a circumference of 56.5 cm. Round your answer to the nearest whole number.

Central Angles

A central angle is an angle whose vertex is at the center of a circle. The sum of the central angles in any circle is 360°.

For a review of *angles*, see page 298.

The part of a circle where a central angle **intercepts** the circle is called an **arc.** The measure of the arc, in degrees, is equal to the measure of the central angle.

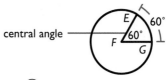

central angle

$\overset{\frown}{EG} = 60°$ and $m\angle EFG = 60°$

 Check It Out

11. Name a central angle of circle Y.
12. What is the measure of $\overset{\frown}{XZ}$?

13. What is the measure of $\overset{\frown}{AB}$?
14. What is the sum of $m\angle AEB$, $m\angle AED$, $m\angle BEC$, and $m\angle CED$?

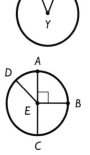

Around the World

Your blood vessels are a network of arteries and veins that carry oxygen to every part of the body and return blood with carbon dioxide to the lungs. Altogether, there are approximately 60,000 miles of blood vessels in the human body.

Just how far is 60,000 miles? The circumference of the earth is about 25,000 miles. If the blood vessels in one human body were placed end to end, approximately how many times would they wrap around the equator? See Hot Solutions for answer.

7·8 CIRCLES

Area of a Circle

To find the area of a circle, you use the formula: Area $=$ pi $\times$ radius^2, or $A = \pi r^2$. As with the area of polygons, the area of a circle is expressed in square units.

For a review of *area* and *square units*, see page 324.

FINDING THE AREA OF A CIRCLE

Find the area of the circle Q to the nearest whole number.

$A = \pi \times 8^2$ • Use the formula $A = \pi r^2$.

  $= 64\pi$ • Square the radius.

  ≈ 200.96 • Multiply by 3.14, or use the calculator

  $\approx 201 \text{ cm}^2$ key for π for a more exact answer.

The area of circle Q is about 201 cm^2.

If you are given the diameter instead of the radius, remember to divide the diameter by two.

✔ Check It Out

15. The diameter of a circle is 14 mm. Express the area of the circle in terms of π. Then multiply and round to the nearest tenth.

16. Use your calculator to find the area of a circle with a diameter of 16 ft. Use the calculator key for π or use $\pi = 3.14$, and round your answer to the nearest square foot.

7·8 EXERCISES

Find the diameter of each circle with the given radius.
1. 6 yd   2. 3.5 cm   3. 2.25 mm

Find the radius of each circle with the given diameter.
4. 20 ft   5. 13 m   6. 8.44 mm

Given the diameter or radius, find the circumference of the circle to the nearest tenth. Use 3.14 for π.
7. $d = 10$ in.   8. $d = 11.2$ m   9. $r = 3$ cm

The circumference of a circle is 88 cm. Find the following, to the nearest tenth.
10. the diameter   11. the radius

Find the measure of each arc of the circle.

12. Arc AB
13. Arc BC
14. Arc ABC

Find the area of each circle with the given radius or diameter. Round to the nearest whole number. Use 3.14 for π.
15. d = 4.5 ft   16. r = 5 in.
17. d = 46 m   18. r = 36 m

19. Which figure has the greater area: a circle with an 8 cm diameter or a square that measures 7 cm on a side?

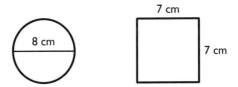

20. Sylvia has a circular glass-top table that measures 42 in. in diameter. When she dropped a jar on the table the glass broke. Now she needs to order replacement glass by the square inch. To the nearest inch, how many square inches of glass will Sylvia need?

What have you learned?

You can use the problems and list of words below to see what you have learned in this chapter. You can find out more about a particular problem or word by referring to the boldfaced topic number (for example, 1•2).

Problem Set

Use this figure for items 1–4. **7•1**

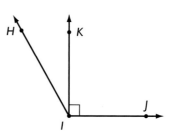

1. Name an angle.

2. Name a ray.

3. What kind of angle is ∠HIJ?

4. Angle *HIK* measures 20°. What is the measure of ∠*HIJ*?

Use this figure for items 5–8. **7•2**

5. What kind of figure is quadrilateral *ABCDE*?
6. Angle *A* measures 108°. What is the measure of ∠*B*?
7. What is the sum of the measures of the angles of quadrilateral *ABCDE*?

For items 8–10 write the letter of the polyhedron, *A*, *B*, or *C*, to match it to its name. **7•2**

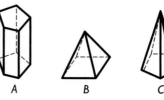

8. pentagonal pyramid
9. square pyramid
10. pentagonal prism

11. What is the perimeter of a regular hexagon that measures 5 in. on a side? **7•4**
12. Find the perimeter of a right triangle whose legs measure 10 mm and 12 mm. **7•4**

13. Find the area of a parallelogram with a base of 19 m and a height of 6 m. **7•5**
14. Find the area of a triangle with a base of 23 ft and a height of 15 ft. **7•5**
15. Find the surface area of a rectangular prism whose length is 32 m, width is 20 m, and height is 5 m. **7•6**
16. What is the surface area of the cylinder if the radius of each base is 3.5 cm and the height is 11 cm? **7•6**
17. Find the volume of a rectangular prism whose dimensions are 12.5 in., 6.3 in., and 2.7 in. **7•7**
18. What is the volume of a cone whose base has a radius of 4 m and whose height is 6.5 m? **7•7**

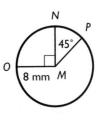

Use circle M to answer items 19 and 20. **7•8**
19. What is the measure of $\overset{\frown}{ONP}$?
20. What is the area of circle M?

---

hot words

WRITE DEFINITIONS FOR THE FOLLOWING WORDS.

angle **7•1**
arc **7•8**
base **7•5**
circle **7•6**
circumference **7•6**
congruent **7•1**
cube **7•2**
cylinder **7•6**
degree **7•1**
diagonal **7•2**
diameter **7•8**
face **7•2**
hexagon **7•2**
legs of a triangle **7•5**
line **7•1**
opposite angle **7•2**

opposite side **7•3**
parallel **7•2**
parallelogram **7•2**
pentagon **7•2**
perimeter **7•4**
perpendicular **7•5**
pi **7•8**
point **7•1**
polygon **7•1**
polyhedron **7•2**
prism **7•2**
pyramid **7•2**
Pythagorean Theorem **7•4**
quadrilateral **7•2**
radius **7•8**
ray **7•1**
rectangular prism **7•2**
reflection **7•3**

regular shape **7•2**
rhombus **7•2**
right angle **7•1**
right triangle **7•4**
rotation **7•3**
segment **7•8**
surface **7•5**
symmetry **7•3**
tetrahedron **7•2**
transformation **7•3**
translation **7•3**
trapezoid **7•2**
triangular prism **7•6**
vertex **7•1**
volume **7•7**

Measurement

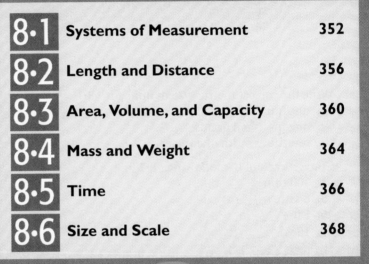

What do you already know?

You can use the problems and list of words below to see what you already know about this chapter. The answers to the problems are in Hot Solutions at the back of the book, and the definitions of the words are in Hot Words at the front of the book. You can find out more about a particular problem or word by referring to the boldfaced topic number (for example, **8•2**).

Problem Set

Write the correct metric system units for: **8•1**
1. one hundredth of a liter
2. one thousand grams
3. one thousandth of a meter

Convert each of the following. **8•2**
4. 100 mm = ___ cm
5. 8 km = ___ m
6. $\frac{1}{2}$ ft = ___ in.
7. 15 yd = ___ ft

Use this rectangle for items 8–13. Round your answer to the nearest whole unit.

Give the perimeter of the rectangle: **8•2**
8. in feet
9. in inches
10. in yards

15 ft

24 ft

Give the area of the rectangle: **8•3**
11. in square feet
12. in square inches
13. in square yards

Convert the following measurements of area, volume, and capacity. **8•3**
14. 2 m^2 = ___ cm^2
15. $\frac{1}{2}\text{ft}^2$ = ___ in.^2
16. 1 yd^3 = ___ ft^3
17. 1 cm^3 = ___ mm^3
18. 1,000 mL = ___ L
19. 2 gal = ___ qt

Imagine that you are going to summer camp and that you have packed your things in this trunk. The packed trunk weighs 96 lb. **8•3, 8•4**

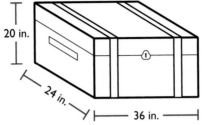

20 in.

24 in.

36 in.

20. What is the volume of the trunk in cubic inches?
21. What is the volume of the trunk in cubic feet?
22. The shipping company can bill you at either of two rates: $5.00 per cubic foot or $.43 per pound. At which rate would you prefer to be billed? Why?
23. When you get to camp, you get a 3 ft × 3 ft × 3 ft space in which to store the trunk. Will it fit?

A team photograph that measures 3 in. × 5 in. needs to be enlarged so it can be displayed in the school trophy case. **8•6**
24. If the photo is enlarged to measure 6 in. × 10 in., what would be the scale factor?
25. If the photo is enlarged by a scale factor of $\frac{3}{2}$, what would be the ratio of the areas of the photos?

WHAT DO YOU KNOW?

CHAPTER 8		
hot **words**	distance **8•2**	round **8•1**
	factors **8•1**	scale factor **8•6**
	fractions **8•1**	side **8•1**
accuracy **8•1**	length **8•2**	similar figures **8•6**
area **8•1**	metric system **8•1**	square **8•1**
customary system **8•1**	power **8•1**	volume **8•3**
	ratio **8•6**	

8·1 Systems of Measurement

If you have ever watched the Olympic Games, you may have noticed that the distances are measured in meters or kilometers, and weights are measured in kilograms. That is because the most common system of measurement in the world is the **metric system.** In the United States, we use the **customary system** of measurement. It will be useful for you to be able to make conversions from one unit of measurement to another within each system, as well as convert units between the two systems.

The Metric and Customary Systems

The metric system of measuring is based on **powers** of ten, such as 10, 100, and 1,000. Converting within the metric system is simple because it is easy to multiply and divide by powers of ten.

Prefixes in the metric system have consistent meanings.

Prefix	Meaning	Example
milli-	one thousandth	1 *milli*liter is 0.001 liter.
centi-	one hundredth	1 *centi*meter is 0.01 meter.
kilo-	one thousand	1 *kilo*gram is 1,000 grams.

BASIC MEASURES

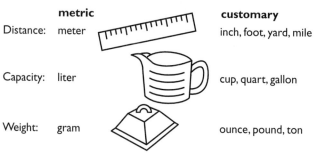

	metric		customary
Distance:	meter		inch, foot, yard, mile
Capacity:	liter		cup, quart, gallon
Weight:	gram		ounce, pound, ton

The customary system of measurement is not based on powers of ten. It is based on numbers like 12 and 16, which have many **factors.** This makes it easy to find, say, $\frac{2}{3}$ ft or $\frac{3}{4}$ lb. While the metric system uses decimals, you will frequently encounter **fractions** in the customary system.

Unfortunately, there are no convenient prefixes as in the metric system, so you will have to memorize the basics: 16 oz = 1 lb; 36 in. = 1 yd; 4 qt = 1 gal; and so on.

Check It Out

1. Which system is based on multiples of 10?
2. Which system uses fractions?

From Boos to Cheers

It took 200 skyjacks two years and 2.5 million rivets to put together the Eiffel Tower. When it was completed in 1899, the art critics of Paris considered it a blight on the landscape. Today, it is one of the most familiar and beloved monuments in the world.

The tower's height, not counting its TV antennas, is 300 meters—that's about 300 yards or 3 football fields. On a clear day, the view can extend for 67 km. Visitors can take elevators to the platforms or climb up the stairs: all 1,652 of them!

Accuracy

Accuracy has to do with both reasonableness and **rounding.** The length of each **side** of the **square** below is measured accurately to the nearest tenth of a meter. But the actual length could be anywhere from 12.15 meters to 12.24 meters. (These are the numbers that all round to 12.2.)

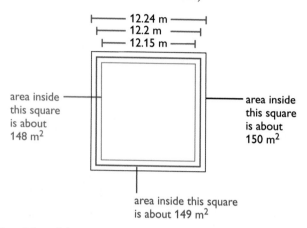

Since the side of the square could really be anywhere between 12.15 m and 12.24 m, the actual **area** may range anywhere between 148 m^2 and 150 m^2. So is it reasonable to square the side $(12.2)^2$ to get an area of 148.84 m^2? No, it isn't. Here is why. The actual length is between 12.15 m and 12.24 m. The area is between 148 m^2 and 150 m^2. Therefore 149 m^2 is reasonable, but the last two digits in 148.84 are meaningless.

Check It Out

Each side of a square measures 6.3 cm (to the nearest tenth).

3. The actual length of the side may range from
   ____ cm to ____ cm.
4. The actual area of the square may range from
   ____ cm^2 to ____ cm^2.

 EXERCISES

What is the meaning of each metric system prefix?
1. milli-
2. centi-
3. kilo-

Write Customary or Metric to identify the system of measurement for the following.
4. ounces and pounds
5. meters and grams
6. feet and miles
7. Which measurement system uses fractions? Which uses decimals?

The measure of the side of the square is given to the nearest tenth. Express the actual area of the square as a range of measurements rounded off to the nearest whole unit.

8·1 EXERCISES

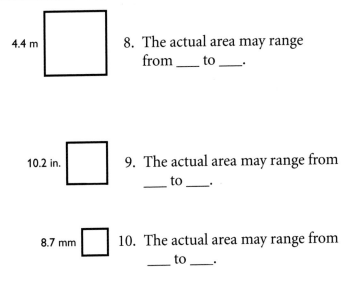

4.4 m

8. The actual area may range from ___ to ___.

10.2 in.

9. The actual area may range from ___ to ___.

8.7 mm

10. The actual area may range from ___ to ___.

8·2 Length and Distance

About What Length?

When you get a feel for "about how long" or "around how far," it's easier to make estimations about **length** and **distance**. Here are some everyday items that will help you keep in mind what metric and customary units mean.

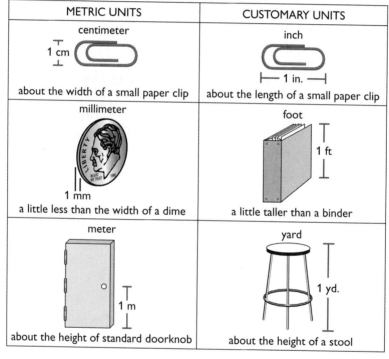

METRIC UNITS	CUSTOMARY UNITS
centimeter	inch
1 cm	1 in.
about the width of a small paper clip	about the length of a small paper clip
millimeter	foot
1 mm	1 ft
a little less than the width of a dime	a little taller than a binder
meter	yard
1 m	1 yd.
about the height of standard doorknob	about the height of a stool

Check It Out

1. Use a metric rule or meter stick to measure common items. Name an item that is about a millimeter; about a centimeter; about a meter.

2. Use a customary rule or yardstick to measure common items. Name an item that is about an inch; about a foot; about a yard.

Metric and Customary Units

When you are calculating length and distance, you may encounter two different *systems of measurement* (p. 352). One is the metric system, and the other is the customary system. The commonly used metric measures for length and distance are millimeter (mm), centimeter (cm), meter (m), and kilometer (km). The customary system uses inch (in.), foot (ft), yard (yd), and mile (mi).

Metric Equivalents

1 km	=	1,000 m	=	100,000 cm	=	1,000,000 mm
0.001 km	=	1 m	=	100 cm	=	1,000 mm
		0.01 m	=	1 cm	=	10 mm
		0.001 m	=	0.1 cm	=	1 mm

Customary Equivalents

1 mi	=	1,760 yd	=	5,280 ft	=	63,360 in.
$\frac{1}{1,760}$ mi	=	1 yd	=	3 ft	=	36 in.
		$\frac{1}{3}$ yd	=	1 ft	=	12 in.
		$\frac{1}{36}$ yd	=	$\frac{1}{12}$ ft	=	1 in.

CHANGING UNITS WITHIN A SYSTEM

How many inches are in $\frac{1}{4}$ mile?

units you have

1 mi = 63,360 in.

conversion factor
for new units

- Find the units you have where they equal 1 on the equivalents chart.

- Find the conversion factor.

$\frac{1}{4} \times 63,360 = 15,840$

- Multiply to get new units.

There are 15,840 inches in $\frac{1}{4}$ mile.

Check It Out

Convert each of the following.

3. 72 in. to ft
4. 10 yd to in.
5. 150 cm to m
6. 50 mm to cm

Conversions Between Systems

Once in a while, you may want to convert between the metric system and the customary system.

CONVERSION TABLE

1 inch	=	25.4 millimeters	1 millimeter	=	0.0394 inch
1 inch	=	2.54 centimeters	1 centimeter	=	0.3937 inch
1 foot	=	0.3048 meter	1 meter	=	3.2808 feet
1 yard	=	0.914 meter	1 meter	=	1.0936 yards
1 mile	=	1.609 kilometers	1 kilometer	=	0.621 mile

To make a conversion, find the listing where the unit you have is 1. Multiply the number of units you have by the conversion factor for the new units.

Your friend in Belgium says he can jump 127 cm. Should you be impressed?

1 cm = 0.3937 in. So 127 × 0.3937 = about 50 in. How far can you jump?

Most of the time you just need to estimate the conversion from one system to the other to get an idea of the size of your item. Round numbers in the conversion table to simplify your thinking. Think that 1 meter is just a little more than 1 yard. 1 inch is between 2 and 3 centimeters. 1 mile is about $1\frac{1}{2}$ kilometers. So now when your friend in Argentina says she caught a fish 60 cm long, you know that the fish is between 20 in. and 30 in. long.

Check It Out

Make exact conversions. Use a calculator, and round to the nearest tenth.

7. Change 10 cm to inches.
8. Change 2 mi to kilometers.

Choose the best estimate.

9. 5 mi is about A. 8 km B. 3 km C. 8 m
10. 100 in. is about
    A. 2.54 cm B. 25 cm C. 254 cm
11. 3 ft is about A. 1 m B. 10 m C. 100 m

EXERCISES

Convert each of the following.

1. 880 yd = ___ mi
2. $\frac{1}{4}$ mi = ___ ft
3. 9 ft = ___ in.
4. 1,500 m = ___ km
5. 17 cm = ___ m
6. 10 mm = ___ cm
7. 5 km = ___ m
8. 0.1 cm = ___ mm
9. 10,560 ft = ___ mi
10. 24 in. = ___ yd

Use a calculator to find the equivalent measurement. Round your answer to the nearest tenth.

11. Change 5 in. to centimeters.
12. Change 3 m to feet.
13. Change 12 ft to meters.
14. Change 25 km to miles.
15. Change 10.5 cm to inches.
16. Change 20 yd to meters.

Choose the best estimate.

17. 12 in. is about
    A. 6 cm
    B. 30 cm
    C. 12 cm
18. 50 mi is about
    A. 80 km
    B. 31 km
    C. 50 km
19. 6 ft is about
    A. 2 m
    B. 3 m
    C. 23 m
20. 1 m is about
    A. 10 yd
    B. 2 ft
    C. 1 yd
21. 6 in. is about
    A. 15.2 mm
    B. 2.4 cm
    C. 152 mm
22. 100 km is about
    A. 161 mi
    B. 6 mi
    C. 62 mi
23. 20 yd is about
    A. 22 km
    B. 18 m
    C. 22 m
24. 10 mm is about
    A. $\frac{1}{2}$ in.
    B. 254 in.
    C. $\frac{1}{10}$ in.
25. 5.5 ft is about
    A. 20 m
    B. 18 m
    C. 2 m

8·3 Area, Volume, and Capacity

Area

Area is the measure of a surface. The walls in your room are surfaces. The large surface of the United States takes up an area of 3,787,319 square miles. The area that the surface of a tire contacts on a wet road makes the difference between skidding and staying in control. Area is given in square units.

Area can be measured in metric units or customary units. Sometimes you might want to convert measurements within a measurement system. You can figure out the conversions by going back to the basic *dimensions* (p. 357). Below is a chart that provides the most common conversions.

<table>
<tr><td colspan="2">Metric</td><td colspan="2">Customary</td></tr>
<tr><td colspan="2">$100 \text{ mm}^2 = 1 \text{ cm}^2$</td><td colspan="2">$144 \text{ in.}^2 = 1 \text{ ft}^2$</td></tr>
<tr><td colspan="2">$10,000 \text{ cm}^2 = 1 \text{ m}^2$</td><td colspan="2">$9 \text{ ft}^2 = 1 \text{ yd}^2$</td></tr>
<tr><td></td><td></td><td colspan="2">$4,840 \text{ yd}^2 = 1 \text{ acre}$</td></tr>
<tr><td></td><td></td><td colspan="2">$640 \text{ acre} = 1 \text{ mi}^2$</td></tr>
</table>

To convert to a new unit, find the 1 for the units you have. Then multiply the units you have by the conversion factor for the new units. If the United States covers an area of about $3,800,000 \text{ mi}^2$, how many acres is it?

$1 \text{ mi}^2 = 640 \text{ acres}$,

so $3,800,000 \text{ mi}^2 \rightarrow 3,800,000 \times 640 = 2,432,000,000 \text{ acres}$

Check It Out

1. Five square centimeters is equal to how many square millimeters?

2. Three square feet is equal to how many square inches?

Volume

Volume is expressed in cubic units. Here are the basic relationships among units of volume.

Metric	Customary
$1{,}000 \text{ mm}^3 = 1 \text{ cm}^3$	$1{,}728 \text{ in.}^3 = 1 \text{ ft}^3$
$1{,}000{,}000 \text{ cm}^3 = 1 \text{ m}^3$	$27 \text{ ft}^3 = 1 \text{ yd}^3$

CONVERTING VOLUME WITHIN A SYSTEM OF MEASUREMENT

Express the volume of the carton in cubic meters.

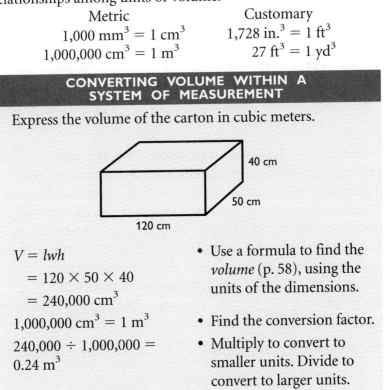

40 cm
50 cm
120 cm

$V = lwh$

$\quad = 120 \times 50 \times 40$

$\quad = 240{,}000 \text{ cm}^3$

$1{,}000{,}000 \text{ cm}^3 = 1 \text{ m}^3$

$240{,}000 \div 1{,}000{,}000 = 0.24 \text{ m}^3$

So the volume of the carton is 0.24 m^3.

- Use a formula to find the *volume* (p. 58), using the units of the dimensions.

- Find the conversion factor.

- Multiply to convert to smaller units. Divide to convert to larger units.

- Include the unit of measurement in your answer.

Check It Out

3. What is the volume of a box that measures 10 in. × 20 in. × 25 in.? Give your answer in cubic feet rounded to the nearest tenth.

4. What is the volume of a box that measures 4 cm on a side? Give your answer in cubic millimeters.

8·3 AREA, VOLUME, AND CAPACITY

Capacity

Capacity is closely related to volume, but there is a difference. A block of wood has volume but no capacity to hold liquid. The capacity of a container is a measure of the volume of liquid it will hold.

<table>
<tr><td align="center">Metric</td><td align="center">Customary</td></tr>
<tr><td align="center">1 liter (L) = 1,000 milliliters (mL)</td><td align="center">8 fl oz = 1 cup (c)</td></tr>
<tr><td align="center">1 L = 1.057 qt</td><td align="center">2 c = 1 pint (pt)</td></tr>
<tr><td></td><td align="center">2 pt = 1 quart (qt)</td></tr>
<tr><td></td><td align="center">4 qt = 1 gallon (gal)</td></tr>
</table>

Note the use of *fluid ounce* (fl oz) in the table. This is to distinguish it from *ounce* (oz) which is a unit of weight (16 oz = 1 lb). The fl oz is a unit of capacity (16 fl oz = 1 pint). There is a connection between oz and fl oz, however. A pint of water weighs about a pound, so a fluid ounce of water weighs about an ounce. For water, as well as for most other liquids used in cooking, *fluid ounce* and *ounce* are equivalent, and the "fl" is sometimes omitted (for example, "8 oz = 1 cup"). To be correct, though, use *ounce* for weight only and *fluid ounce* for capacity. For liquids that weigh considerably more or less than water, the difference is significant.

In the metric system, the basic units of capacity are related.

1 liter (L) = 1,000 milliliters (mL)
Think of a liter as having the capacity of about a quart.
1 L = 1.057 qt

Gasoline is priced at $0.39/L. How much is that per gallon? There are 4 quarts in a gallon, so there are 4 × 1.057 or 4.228 liters in a gallon. So a gallon costs $0.39 × 4.228 or $1.649 per gallon.

Check It Out

You have a 1-pint container to use to fill other containers with water. How many pints do you need to fill a container with capacity:

5. 1 gallon
6. 1 cup

EXERCISES

Tell whether the unit of measure is used to express distance, area, or volume.
1. square centimeter
2. mile
3. cubic meter
4. kilometer

For items 5–7, express the volume of this box:

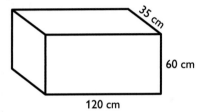

35 cm

60 cm

120 cm

5. in cubic centimeters
6. in cubic millimeters
7. in cubic meters

Convert each of the following.
8. 1 L = ___ mL
9. 4 qt = ___ gal
10. 1.057 qt = ___ L
11. 500 mL = ___ L
12. 2 gal = ___ c
13. 1 qt = ___ fl. oz
14. 3 qt = ___ gal
15. 8 c = ___ qt
16. 5 c = ___ qt
17. 16 fl. oz = ___ c
18. 5,000 mL = ___ L

19. When Sujey said, "It holds about 2 liters," was she talking about a bathtub, a bottle of cola, or a paper cup?
20. When Pei said, "It holds about 250 gallons," was he pointing to an oil tank, a milk container, or an in-ground swimming pool?

8·3 EXERCISES

8·4 Mass and Weight

Technically, mass and weight are different. Mass is the amount of substance you have. Weight is the pull of gravity on the amount of substance. On Earth, mass and weight are equal at sea level and about equal at other elevations. But on the moon, mass and weight can be quite different. Your mass would be the same on the moon as it is here on Earth. But, if you weigh 100 pounds on Earth, you would weigh about $16\frac{2}{3}$ pounds on the moon. That is because the gravitational pull of the moon is only $\frac{1}{6}$ that of the Earth.

The customary system measures weight. The metric system measures mass.

Metric	Customary
1 kg = 1,000 g = 1,000,000 mg	1 T = 2,000 lb = 32,000 oz
0.001 kg = 1 g = 1,000 mg	0.0005 T = 1 lb = 16 oz
0.000001 kg = 0.001 g = 1 mg	0.0625 lb = 1 oz

$$1\ lb \approx 0.4536\ kg$$
$$1\ kg \approx 2.205\ lb$$

To convert from one unit of mass or weight to another, first find the 1 for the units you have in the equivalents chart. Then multiply the number of units you have by the conversion factor for the new units.

If you have 64 oz of peanut butter, how many pounds do you have? 1 oz = 0.0625 lb, so 64 oz = 64 × 0.0625 lb = 4 lb. You have 4 pounds of peanut butter.

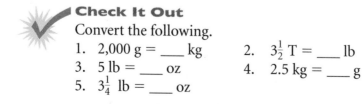

Check It Out

Convert the following.

1. 2,000 g = ____ kg
2. $3\frac{1}{2}$ T = ____ lb
3. 5 lb = ____ oz
4. 2.5 kg = ____ g
5. $3\frac{1}{4}$ lb = ____ oz

8·4 EXERCISES

Convert the following measurements.

1. 7 kg = ___ mg
2. 500 mg = ___ g
3. 1,000 lb = ___ T
4. 8 oz = ___ lb
5. 200,000 mg = ___ kg
6. 2 T = ___ lb
7. 2,000 mg = ___ kg
8. 6 kg = ___ lb
9. 15 lb = ___ kg
10. 100 lb = ___ oz
11. $\frac{1}{2}$ lb = ___ oz
12. 5 g = ___ mg
13. 4,000 g = ___ kg
14. 48 oz = ___ lb
15. 2.5 g = ___ mg
16. 2 T = ___ oz

17. The weight of a baby born in Zaire was recorded as 2.7 kg. If that baby had been born in the United States, what weight would have been recorded in pounds?
18. Nirupa's scale shows her weight in both pounds and kilograms. She weighed 95 lb one morning. How many kilograms was that?
19. Coffee is on sale at Slonim's for $3.99 per lb. The same brand is advertised at Harrow's for $3.99 per 0.5 kg. Which store has the better buy?
20. A baker needs 15 lb of flour to make bread. How many 2-kg bags of flour will be needed?

Poor SID

SID is a crash-test dummy. After a crash, SID goes to the laboratory for a readjustment of sensors and perhaps a replacement head or other body parts. Because of the forces at work when a car crashes, body parts weigh as much as 20 times their normal weight.

The weight of a body changes during a crash. Does the mass of the body also change? See Hot Solutions for the answer.

8·4 EXERCISES

 **Time**

Time measures the interval between two or more events. You can measure time with a very short unit—a second—or a very long unit—a millennium—and many units in between.

1,000,000 seconds before 12:00 A.M., January 1, 2000 is 10:13:20 A.M., December 20, 1999. 1,000,000 hours before 12:00 A.M., January 1, 2000 is 8:00 A.M., December 8, 1885.

60 seconds (sec) = 1 minute (min)	365 da = 1 year (yr)
60 min = 1 hour (hr)	10 yr = 1 decade
24 hr = 1 day (da)	100 yr = 1 century
7 da = 1 week (wk)	1,000 yr = 1 millennium

Working with Time Units

Like other kinds of measurement, you can convert one unit of time to another, using the information in the table above.

Hulleah is 13 years old. Her age in months is 13 × 12, or 156 months.

Leap Years

Every four years, February has an extra day. These 366 day years are called *leap years*. Leap years are divisible by 4, but not by 100. However, years that are divisible by 400 are leap years. The year 1996 is a leap year, but the year 1900 is not. The year 2000 is also a leap year.

 Check It Out

1. How many months old will you be on your twenty-first birthday?
2. What date will it be 5,000 days from January 1, 2000?

EXERCISES

Convert each of the following units.
1. 2 da = ___ hr
2. 90 min = ___ hr
3. 1 yr = ___ da
4. 200 yr = ___ centuries
5. 5 min = ___ sec
6. 30 centuries = ___ millennia
7. How many days are there in three 365-day years?
8. How many days are there in four years (including one leap year)?
9. How many hours are there in a week?
10. How many years old will you be when you have lived for 6,939 days?

The World's Largest Reptile

Would it surprise you to learn that the world's largest reptile is a turtle? The leatherback turtle can weigh as much as 2,000 pounds. By comparison, an adult male crocodile weighs about 1,000 pounds.

The leatherback has existed in its current form for over 20 million years, but this prehistoric giant is now endangered. If after 20 million years of existence the leatherback was to become extinct, how many times longer than Homo sapiens will it have existed? Assume Homo sapiens have been around 4,000 millennia. See Hot Solutions for answer.

8·5 EXERCISES

8·6 Size and Scale

Similar Figures

Similar figures are figures that have exactly the same shape. When two figures are similar, one may be larger than the other.

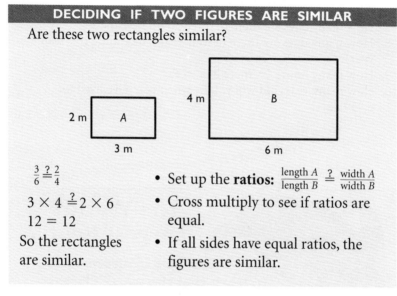

DECIDING IF TWO FIGURES ARE SIMILAR

Are these two rectangles similar?

$\frac{3}{6} \stackrel{?}{=} \frac{2}{4}$

$3 \times 4 \stackrel{?}{=} 2 \times 6$

$12 = 12$

So the rectangles are similar.

- Set up the **ratios:** $\frac{\text{length } A}{\text{length } B} \stackrel{?}{=} \frac{\text{width } A}{\text{width } B}$
- Cross multiply to see if ratios are equal.
- If all sides have equal ratios, the figures are similar.

Check It Out

1. Which figures are similar to the shaded figure?

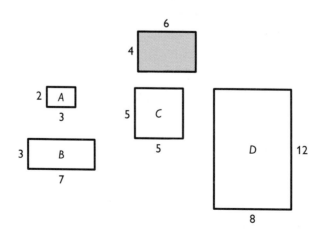

Scale Factors

A **scale factor** indicates the ratio of sizes of two similar figures.

Triangle A is similar to triangle B. $\triangle B$ is 3 times larger than $\triangle A$. The scale factor is 3.

FINDING THE SCALE FACTOR

What is the scale factor for these similar pentagons?

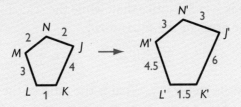

- Decide which figure is the "original figure."

$\dfrac{K'J'}{KJ} = \dfrac{6}{4}$

- Make a ratio of corresponding sides:

$$\frac{\text{new figure}}{\text{original figure}}$$

$= \dfrac{3}{2}$

- Reduce, if possible.

The scale factor of the similar pentagons is $\dfrac{3}{2}$.

When a figure is enlarged, the scale factor is greater than 1. When two similar figures are identical in size, the scale factor is equal to 1. When a figure is reduced, the scale factor is less than 1.

Check It Out

What is the scale factor for each pair of similar figures?

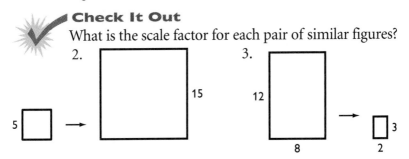

2.

3.

8·6 SIZE AND SCALE

Scale Factors and Area

Scale factor refers to a ratio of the lengths of the sides of two similar figures only, not the ratio of their areas.

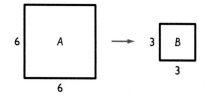

For the squares above, the scale factor is $\frac{1}{2}$ because the ratio of their sides is $\frac{3}{6} = \frac{1}{2}$. Notice that, while the scale factor is $\frac{1}{2}$, the ratio of the areas of the squares is $\frac{1}{4}$.

$$\frac{\text{Area of } B}{\text{Area of } A} = \frac{3^2}{6^2} = \frac{9}{36} = \frac{1}{4}$$

The scale factor is $\frac{1}{3}$. What is the ratio of the areas?

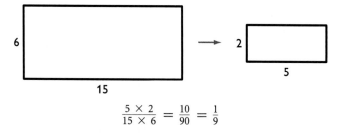

$$\frac{5 \times 2}{15 \times 6} = \frac{10}{90} = \frac{1}{9}$$

The ratio of the areas is $\frac{1}{9}$.

In general, the ratio of the areas of two similar figures is the *square* of the scale factor.

Check It Out

4. The scale factor for two similar figures is $\frac{1}{4}$. What is the ratio of the areas?

5. A 4 in. × 6 in. photograph is enlarged by a scale factor of 2. How big is the enlarged photo?

 **EXERCISES**

Give the scale factor for each pair of similar figures.

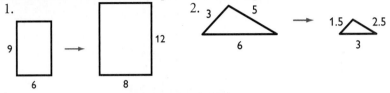

1.

2.

Given the scale factors, find the missing dimensions for the sides of the figures.

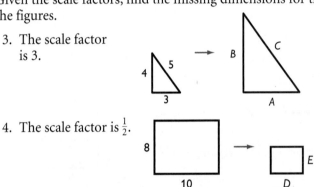

3. The scale factor is 3.

4. The scale factor is $\frac{1}{2}$.

5. The scale used for a drawing of a playhouse is 1 in. = 2 ft. An area of 1 in.^{2} on the drawing stands for how much of the area of the playroom floor?

Imagine you are making photocopies of a birth certificate that measures 4 in. by 5 in.

6. Enlarge the birth certificate by a scale factor of 2. What are the dimensions of the enlargement?

7. Reduce the birth certificate by a scale factor of $\frac{1}{2}$. What are the dimensions of the reduction?

A map is drawn to the scale 1 cm = 2 km.

8. If the distance between the school and the post office measures 5 cm on the map, what is the actual distance between them?

9. If the actual distance between the library and the theater is 5 km, how far apart do they appear on the map?

10. The highway that runs from one end of the city to the opposite end measures 3.5 cm on the map. What is the actual distance of this section of the highway?

What have you learned?

You can use the problems and list of words below to see what you have learned in this chapter. You can find out more about a particular problem or word by referring to the boldfaced topic bold number (for example, **8•2**).

Problem Set

Write the correct metric system units for: **8•1**
1. one thousandth of a meter
2. one hundredth of a liter
3. one thousand grams

Convert each of the following. **8•2**
4. 350 mm = ___ m
5. 0.07 m = ___ mm
6. 6 in. = ___ yd

Use this rectangle for items 8–13. Round your answer to the nearest whole unit.

Give the perimeter of the rectangle: **8•2**

7. in feet
8. in inches

12 ft

30 ft

Give the area of the rectangle: **8•3**
9. in square feet
10. in square inches

Convert the following measurements of area, volume, and capacity. **8•3**
11. $10\ m^2 =$ ___ cm^2
12. $4\ ft^2 =$ ___ $in.^2$
13. $3\ yd^3 =$ ___ ft^3
14. $4\ cm^3 =$ ___ mm^3
15. 3,000 mL = ___ L
16. 6 gal = ___ qt

A family photograph that measures 4 in. by 6 in. needs to be enlarged so it can be displayed in a new frame. **8•6**

17. If the photo is enlarged to measure 12 in. by 18 in., what would be the scale factor?

18. If the photo is enlarged by a scale factor of $\frac{3}{2}$, what would be the ratio of the areas of the photos?

Imagine that you are going to summer camp and that you have packed your things in this trunk. The packed trunk weighs 106 lb. **8•3, 8•4**

24 in.

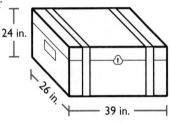

26 in.

39 in.

19. What is the volume of the trunk in inches?

20. The shipping company can bill you at either of two rates: $4.00 per cubic foot or $0.53 per pound. At which rate would you prefer to be billed? Why?

---

hot words

WRITE DEFINITIONS FOR THE FOLLOWING WORDS.

accuracy **8•1**
area **8•3**
customary system **8•1**

distance **8•2**
factor **8•1**
fraction **8•1**
length **8•2**
metric system **8•1**
power **8•1**
ratio **8•6**

round **8•1**
scale factor **8•6**
side **8•1**
similar figures **8•6**
square **8•1**
volume **8•3**

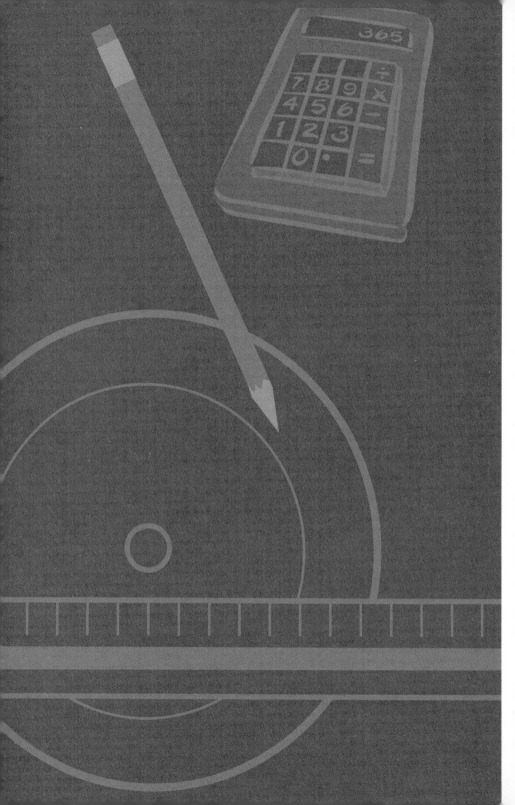

Tools

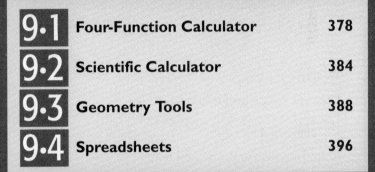

What do you already know?

You can use the problems and list of words below to see what you already know about this chapter. The answers to the problems are in Hot Solutions at the back of the book, and the definitions of the words are in Hot Words at the front of the book. You can find out more about a particular problem or word by referring to the boldfaced topic number (for example, **9•2**).

Problem Set

Use your calculator for items 1–6. **9•1**

1. $40 + 7 \times 5 + 4^2$
2. 300% of 450

Round answers to the nearest tenth.

3. $8 + 3.75 \times 5^2 + 15$
4. $62 + (-30) \div 0.5 - 12.25$

5. Find the perimeter of rectangle *ABCD*.
6. Find the area of rectangle *ABCD*.

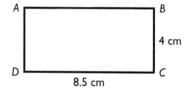

Use a scientific calculator for items 7–12. Round decimal answers to the nearest hundredth. **9•2**

7. 5.5^3
8. Find the reciprocal of 8.
9. Find the square of 12.4.
10. Find the square root of 4.5.
11. $(2 \times 10^3) \times (9 \times 10^2)$
12. $7.5 \times (6 \times 3.75)$

13. What is the measure of ∠*VRT*? **9•3**
14. What is the measure of ∠*SRV*? **9•3**
15. What is the measure of ∠*SRT*? **9•3**
16. Does ray *RT* divide ∠*SRV* into two equal angles? **9•3**

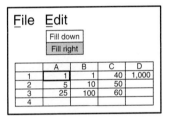

17. What are the basic construction tools in geometry? **9•3**

For items 18–20, refer to the spreadsheet. **9•4**

	A	B	C	D
1	1	1	40	1,000
2	5	10	50	
3	25	100	60	
4				

File Edit
Fill down
Fill right

18. Name the cell holding 1,000.
19. A formula for cell B2 is B1 × 10. Name another formula for cell B2.
20. Cell D1 contains the number 1,000 and no formula. After using the command *fill down*, what number will be in cell D3?

CHAPTER 9

hot **words**

	distance **9•3**	power **9•2**
	factorial **9•2**	radius **9•1**
	formula **9•4**	ray **9•3**
angle **9•3**	horizontal **9•4**	reciprocal **9•2**
arc **9•3**	negative number **9•1**	root **9•2**
cell **9•4**		row **9•4**
circle **9•1**	parentheses **9•2**	spreadsheet **9•4**
column **9•4**	percent **9•1**	square **9•2**
cube **9•2**	perimeter **9•4**	square root **9•1**
cube root **9•2**	pi **9•1**	vertex **9•3**
decimal **9•1**	point **9•3**	vertical **9•4**

9.1 Four-Function Calculator

People use calculators to make mathematical tasks easier. You might have seen your parents balance their checkbooks using a calculator. But a calculator is not always the fastest way to do a mathematical task. If your answer does not need to be exact, it might be faster to estimate. Sometimes you can do the problem in your head quickly, or a pencil and paper might be a better method. Calculators are particularly helpful for problems with many numbers or with numbers that have many digits.

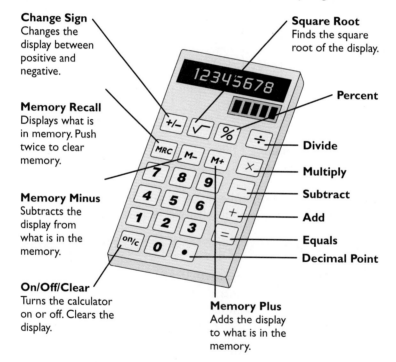

Change Sign
Changes the display between positive and negative.

Square Root
Finds the square root of the display.

Percent

Memory Recall
Displays what is in memory. Push twice to clear memory.

Divide

Multiply

Subtract

Memory Minus
Subtracts the display from what is in the memory.

Add

Equals

Decimal Point

On/Off/Clear
Turns the calculator on or off. Clears the display.

Memory Plus
Adds the display to what is in the memory.

A calculator only gives you the answer to the problem you enter. Always have an estimate of the answer you expect. Then you can compare the calculator answer to your estimate to be sure you entered the problem correctly.

Basic Operations

Adding, subtracting, multiplying, and dividing are fairly straightforward.

Operation	Problem	Calculator Keys	Display
Addition	$10.5 + 39$	10.5 $\boxed{+}$ 39 $\boxed{=}$	49.5
Subtraction	$40 - 51$	40 $\boxed{-}$ 51 $\boxed{=}$	−11.
Multiplication	20.5×4	20.5 $\boxed{\times}$ 4 $\boxed{=}$	82.
Division	$12 \div 40$	12 $\boxed{\div}$ 40 $\boxed{=}$	0.3

Negative Numbers

To enter a **negative number** into your calculator, you press $\boxed{+/-}$ after you enter the number.

Problem	Calculator Keys	Display
$-15 + 10$	15 $\boxed{+/-}$ $\boxed{+}$ 10 $\boxed{=}$	−5.
$50 - (-32)$	50 $\boxed{-}$ 32 $\boxed{+/-}$ $\boxed{=}$	82.
-9×8	9 $\boxed{+/-}$ $\boxed{\times}$ 8 $\boxed{=}$	−72.
$-20 \div (-4)$	20 $\boxed{+/-}$ $\boxed{\div}$ 4 $\boxed{+/-}$ $\boxed{=}$	5.

Check It Out

Find each answer on a calculator.

1. $11.6 + 4.2$
2. $45.4 - 13.9$
3. $20 \times (-1.5)$
4. $-24 \div 0.5$

9•1 FOUR-FUNCTION CALCULATOR

Memory

For complex or multi-step problems, use memory. You operate memory with three special keys. The way many calculators operate is shown below. If yours does not work this way, check the instructions which came with your calculator.

Key | **Function**

MRC — One push displays (recalls) what is in memory. Push twice to clear memory.

M+ — Adds display to what is in memory.

M− — Subtracts display from what is in memory.

When calculator memory contains something other than zero, the display will show $\boxed{^M \qquad}$ along with whatever number the display currently shows. What you do on your calculator does not change memory unless you use the special memory keys.

To solve $10 + 55 + 26 \times 2 + 60 - 4^2$ you could use the following keystrokes to do the problem with your calculator:

Keystrokes	Display
MRC MRC C	0.
4 × 4 M−	^{M} 16.
26 × 2 M+	^{M} 52.
10 + 55 M+	^{M} 65.
60 M+	^{M} 60.
MRC	161.

Your answer is 161. Notice the use of *order of operations* (p. 78).

Check It Out

Use memory to find each answer.

5. $5 \times 10 - 18 \times 3 + 8^2$
6. $-40 + 5^2 - (-14) \times 6$
7. $6^3 \times 4 + (-18) \times 20 + (-50)$
8. $20^2 + 30 \times (-2) - (-60)$

Special Keys

Some calculators have keys with special functions to save time.

Key **Function**

$\sqrt{x}$ Finds the **square root** of the display.

% Changes display to the decimal expression of a **percent**.

π Automatically enters **pi** to as many places as your calculator holds.

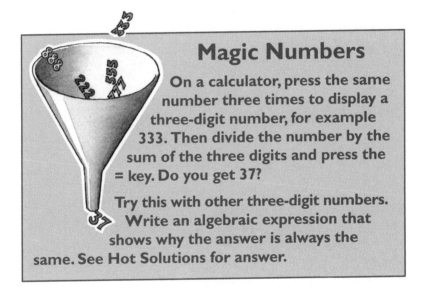

Magic Numbers

On a calculator, press the same number three times to display a three-digit number, for example 333. Then divide the number by the sum of the three digits and press the = key. Do you get 37?

Try this with other three-digit numbers. Write an algebraic expression that shows why the answer is always the same. See Hot Solutions for answer.

9·1 FOUR-FUNCTION CALCULATOR

The % and π keys save you time by saving you keystrokes. The √ key allows you to find square roots more precisely, something difficult to do by hand. See how they work in the examples below.

Problem: $10 + \sqrt{144}$

Keystrokes: 10 + 144 √ =

Final display: 22.

If you try to take a square root of a negative number, your calculator will display an error message, such as 64 +/- √ E 5. . There is no square root of -64, because no number times itself can give a negative number.

Problem: Find 40% of 50.

Keystrokes: 50 × 40 %

Final display: 20.

The % key only changes a percent to its decimal form. If you know how to convert percents to decimals, you probably will not use the % key much.

Problem: Find the area of a **circle** with **radius** 2.
(Use formula $A = \pi r^2$.)

Keystrokes: π × 2 × 2 =

Final display: 12.57

If your calculator does not have the π key, you can use 3.14 or 3.1416 as an approximation for π.

Check It Out

9. Without using the calculator, tell what the display would be if you entered: 12 M+ 4 × 2 + MRC = .

10. Use memory functions to find the answer to $160 - 8^2 \times (-6)$.

11. Find the square root of 196.

12. Find 25% of 450.

9·1 EXERCISES

Find the value of each expression, using your calculator.

1. $15.6 + 22.4$ 2. $45.61 - 20.8$ 3. $-16.5 - 5.6$

4. $10 \times 45 \times 30$ 5. $-5 + 60 \times (-9)$ 6. $50 - 12 \times 20$

7. $\sqrt{81} - 16$ 8. $-10 + \sqrt{225}$ 9. $12 \div 20 + 11$

10. $12 \div (-20)$ 11. 20% of 350   12. 120% of 200

13. $216 - \sqrt{484}$ 14. $\sqrt{324} \div 2.5 + 8.15$ 15. $5 \times \sqrt{49} + 1.73$

Use a calculator to answer items 17–25.
16. Find the area if $x = 2.5$ cm.
17. Find the perimeter if $x = 4.2$ cm.

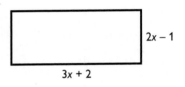

18. Find the area if $a = 1.5$ in.
19. Find the circumference if $a = 3.1$ in.

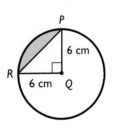

20. Find the area of $\triangle RQP$.
21. Find the perimeter of $\triangle RQP$. (*Remember:* $a^2 + b^2 = c^2$.)
22. Find the circumference of circle Q.
23. Find the area of circle Q.
24. Find the length of line segment RP.
25. Find the area of the shaded part of circle Q.

9·2 Scientific Calculator

Every mathematician and scientist has a scientific calculator to help quickly and accurately solve complex equations. Scientific calculators vary widely, some with a few functions and others with many functions. Some calculators can even be programmed with functions of your choosing. The calculator below shows functions you might find on your scientific calculator.

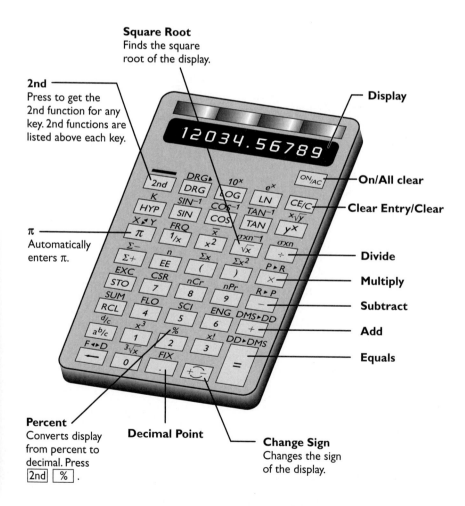

Square Root
Finds the square root of the display.

2nd
Press to get the 2nd function for any key. 2nd functions are listed above each key.

Display

On/All clear

Clear Entry/Clear

π
Automatically enters π.

Divide

Multiply

Subtract

Add

Equals

Percent
Converts display from percent to decimal. Press 2nd % .

Decimal Point

Change Sign
Changes the sign of the display.

Frequently Used Functions

Since each scientific calculator is set up differently, your calculator may not work exactly as below. These keystrokes work with the calculator illustrated on page 384. Use the reference book or card that came with your calculator to perform similar functions. See the index to find more information about the mathematics here.

Function	Problem	Keystrokes
Cube Root $\sqrt[3]{x}$ Finds the cube root of the display.	$\sqrt[3]{64}$	64 [2nd] $\sqrt[3]{x}$ [4.]
Cube x^3 Finds the cube of the display.	5^3	5 [2nd] x^3 [125.]
Factorial $x!$ Finds the factorial of the display.	$5!$	5 [2nd] $x!$ [120.]
Fix number of **decimal places** [FIX] Rounds display to number of places you determine.	Round 3.729 to the hundredths place.	3.729 [2nd] [FIX] 2 [3.73]
Parentheses [( ] [ )] Use to group calculations	$8 \times (7 + 2)$	8 [×] [( ] 7 [+] 2 [ )] [=] [72.]
Powers y^x Finds the x power of the display	12^4	12 y^x 4 [=] [20736.]
Powers of ten 10^x Raises ten to the power displayed.	10^3	3 [2nd] 10^x [1000.]

Function	Problem	Keystrokes
Reciprocal $\boxed{1/x}$ Finds the reciprocal of the display.	Find the reciprocal of 10.	10 $\boxed{1/x}$ $\boxed{\qquad 0.1}$
Roots $\boxed{\sqrt[x]{y}}$ Finds the x root of the display.	$\sqrt[4]{1296}$	1,296 $\boxed{\text{2nd}}$ $\boxed{\sqrt[x]{y}}$ 4 $\boxed{=}$ $\boxed{\qquad 6.}$
Square $\boxed{x^2}$ Finds the square of the display.	9^2	9 $\boxed{x^2}$ $\boxed{\qquad 81.}$

Check It Out

Use your calculator to find the following
1. $\sqrt[3]{91.125}$ 2. 7^3
3. $7!$ 4. 9^4
5. $\sqrt[5]{243}$

Use your calculator to find the following to the
nearest thousandth.
6. $6 \times (21 - 3) \div (2 \times 5)$
7. the reciprocal of 2
8. 19^2
9. $\sqrt[3]{512} \times 4^4 + \sqrt{400}$
10. $(6^2 - 9^3 + \sqrt[4]{625}) \div 2$

 **EXERCISES**

Use a scientific calculator to find the following.
1. 18^2 2. 9^3 3. 12^3 4. 2.5^2

5–9. Give your answer to the nearest hundredth.
5. 3π 6. $\frac{30}{\pi}$ 7. $\frac{1}{5}$
8. $\frac{2}{\pi}$ 9. $(10 + 4.1)^2 + 4$
10. $12 - (20 \div 2.5)$ 11. $2! \times 3!$ 12. $8! \div 3!$
13. $5! + 6!$ 14. $\sqrt[4]{1296}$
15. reciprocal of 40

Calculator Alphabet

You probably think calculators are only useful for doing arithmetic. But you can also send "secret" messages with them—if you know the calculator alphabet. Try this. Enter the number 0.7734 in your calculator and turn it upside down. What word appears in the display?

Each of the calculator's numerical keys can be used to display a letter. You can enter:

- 8 to display a B.
- 6 to display a g.
- 4 to display an h.
- 7 to display an L.
- 5 to display an S.

- 3 to display an E.
- 9 to display a G or b.
- 1 to display an I.
- 0 to display an O.
- 2 to display a Z.

See how many words you can display with your calculator. Remember, you will be turning the calculator upside down to read the words, so enter the numbers in reverse order.

9.3 Geometry Tools

Ruler

If you need to measure the dimensions of an object, or if you need to measure reasonably short **distances,** use a ruler.

A metric ruler

A customary ruler

To get an accurate measure, be sure one end of the item being measured lines up with zero on your ruler.

The pencil below is measured first to the nearest tenth of a centimeter and then to the nearest eighth of an inch.

The pencil is about 9.8 cm long.

No. 2

The pencil is about $3\frac{7}{8}$ in. long.

Check It Out

Use your ruler. Measure each line segment to the nearest tenth of a centimeter or the nearest eighth of an inch.

1. ───────────
2. ───────────────
3. ────────────────────
4. ──────────

Protractor

Measure **angles** with a *protractor.* There are many different protractors. The key is to find the point on each protractor to which you align the **vertex** of the angle.

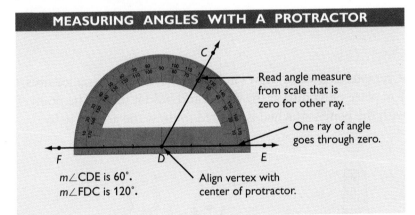

MEASURING ANGLES WITH A PROTRACTOR

Read angle measure from scale that is zero for other ray.

One ray of angle goes through zero.

Align vertex with center of protractor.

$m\angle CDE$ is 60°.
$m\angle FDC$ is 120°.

To draw an angle with a given measure, draw one **ray** first and position the center of the protractor at the endpoint. Then make a dot at the desired measure (45°, in this example).

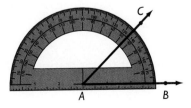

Connect *A* and *C*. Then $\angle BAC$ is a 45° angle.

Check It Out

Measure each angle to the nearest degree using your protractor.

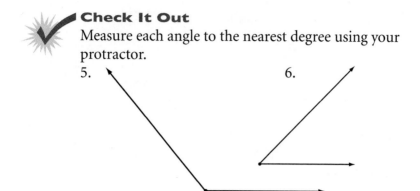

5.

6.

Compass

A *compass* is used to draw circles or parts of circles, called **arcs**. You place one **point** at the center and hold it there. The point with the pencil attached is pivoted to draw the arc or circle.

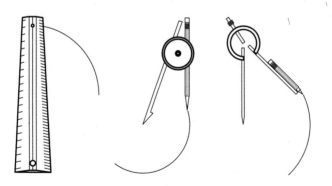

The distance between the point that is stationary (the center) and the pencil is the radius. Some compasses allow you to set the radius exactly.

For a review of *circles*, see page 340.

To draw a circle with a radius of $1\frac{1}{2}$ in., set the distance between the stationary point of your compass and the pencil at $1\frac{1}{2}$ in. Draw a circle.

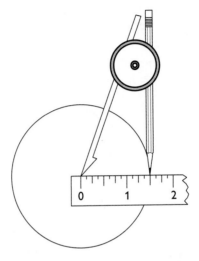

Check It Out

7. Draw a circle with a radius of 4 in. or 10.2 cm.
8. Draw a circle with a radius of 4 cm or 1.6 in.
9. Draw a circle with a radius of 3 cm or 1.2 in.
10. Draw a circle with a radius of 3 in. or 7.6 cm.

Construction Problem

A construction is a drawing problem in geometry that permits the use of only the straightedge and the compass. When you make a construction using the straightedge and the compass, you have to use what you know about geometry.

Follow the step-by-step directions below to inscribe an equilateral triangle in a circle.

- Draw a circle with center K.
- Draw a diameter ($\overline{SJ}$).
- Using S as a center and $\overline{SK}$ as a radius, draw an arc intersecting the circle at L and P.
- Connect L, P, and J to form the triangle.

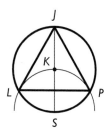

You can create a more complex design by inscribing another triangle in your circle using J as a center for drawing another intersecting arc.

Once you have the framework, you can fill in different sections to come up with a variety of designs based on constructions.

Check It Out

11. Draw the framework based on two triangles inscribed in a circle. Fill in sections to copy the design below.

12. Create your own design based on one or two triangles inscribed in a circle.

Mandalas

A mandala is a design which consists of geometric shapes and symbols that are meaningful to the artist. The word *mandala* means "circle" in Sanskrit, and the mandala design is usually contained within a circle.

In Hinduism and Buddhism, mandalas are used as aids to meditation and often incorporate symbols for the gods or the universe. Western artists create mandalas to symbolize their own lives or the lives of famous people. Within the geometric patterns, symbols for animals, the elements (earth, wind, fire, and water), the sun and stars, as well as personal symbols, appear frequently.

9·3 GEOMETRY TOOLS

9.3 EXERCISES

Using a ruler, measure the length of each side of $\triangle ABC$. Give your answer in inches or centimeters, rounded to the nearest $\frac{1}{8}$ in. or $\frac{1}{10}$ cm.

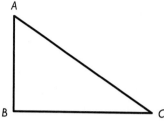

1. AB
2. BC
3. AC

Using a protractor, measure each angle in $\triangle ABC$.

4. $\angle C$ 　　　　　 5. $\angle B$ 　　　　　 6. $\angle A$

7. What is the sum of the measures of the interior angles of any triangle? Explain.

Match each function with the correct tool.

Function	Tool
8. Draw circles or arcs	A. protractor
9. Measure distance	B. compass
10. Measure angles	C. ruler

Write the measures of the following angles.

11. $\angle GFH$
12. $\angle HFJ$
13. $\angle JFG$
14. $\angle HFI$
15. $\angle IFJ$

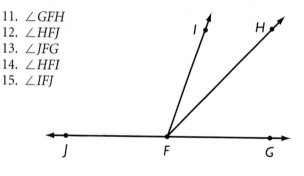

Write the measures of the following angles.
16. $\angle NML$
17. $\angle MLK$
18. $\angle KNM$
19. $\angle LKN$

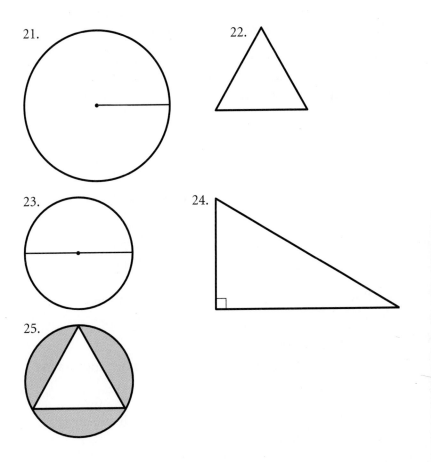

20. Use a protractor to copy $\angle MLK$.

Using a ruler, protractor, and compass, copy the figures below.

21.

22.

23.

24.

25.

9.4 Spreadsheets

What Is a Spreadsheet?

People have used **spreadsheets** as tools to keep track of information, such as finances, for a long time. Spreadsheets were paper-and-pencil math tools before becoming computerized. You may be familiar with computer spreadsheet programs.

A spreadsheet is a computer tool where information is arranged into **cells** within a grid and calculations are performed within the cells. When one cell is changed, all other cells that depend on it automatically change.

Spreadsheets are organized into **rows** and **columns.** Rows are **horizontal** and are numbered. Columns are **vertical** and are named by capital letters. The cells are named for their rows and columns.

```
 File   Edit

         A     B     C     D
   1     1     3     1
   2     2     6     4
   3     3     9     9
   4     4    12    16
   5     5    15    25
   6
   7
   8
```

The cell A3 is in Column A, Row 3. In this spreadsheet, there is a 3 in cell A3.

Check It Out

In the spreadsheet above, what number appears in each cell?

1. A5   2. B2   3. C2

Answer the following statements with true or false.

4. A column is vertical.
5. A row is labeled with letters.

Spreadsheet Formulas

A cell can contain a number, or it may contain the information it needs to generate a number. A **formula** generates a number dependent on other cells in the spreadsheet. The way the formulas are written depends on the particular spreadsheet computer software you are using. You enter a formula and the value generated shows, not the formula.

CREATING A SPREADSHEET FORMULA

	A	B	C	D
1	Item	Price	Qty	Total
2	sweater	$30	2	$60
3	pants	$18	2	
4	shirt	$10	4	
5				
6				

Express the value of the cell in relationship to other cells.

$$Total = Price \times Qty$$
$$D2 = B2 \times C2$$

If you change the value of a cell and a formula depends on it, the result of the formula will change.

> In the spreadsheet above, if you entered 3 sweaters instead of 2 ($C2 = 3$), the Total column would automatically change to $90.

 Check It Out

Use the spreadsheet above. If the Total is always figured the same way, write the formula for:

6. D3
7. D4

8. What is the price of one shirt?
9. What is the total spent on sweaters?

Fill Down and Fill Right

Now that you know the basics, let's look at some ways to make spreadsheets do even more of the work for you. *Fill down* and *fill right* are two spreadsheet commands that can save you a lot of time and effort.

To use *fill down,* select a portion of a column. *Fill down* will take the top cell that has been selected and copy it into the lower cells. If the top cell in the selected range contains a number, such as 5, *fill down* will generate a column containing all 5s.

If the top cell of the selected range contains a formula, the *fill down* feature will automatically adjust the formula as you go from cell to cell.

The selected column is highlighted.

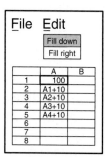

The spreadsheet fills the column and adjusts the formula.

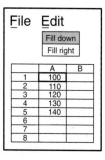

These are the values that actually appear.

Fill right works in a similar manner, except it goes across, copying the leftmost cell of the selected range in a row.

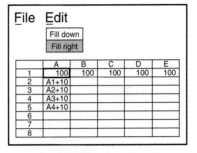

	A	B	C	D	E
1	100				
2	A1+10				
3	A2+10				
4	A3+10				
5	A4+10				
6					
7					
8					

	A	B	C	D	E
1	100	100	100	100	100
2	A1+10				
3	A2+10				
4	A3+10				
5	A4+10				
6					
7					
8					

Row 1 is selected.   The 100 fills to the right.

If you select A1 to E1 and fill right, you will get all 100s.
If you select A2 to E2 and fill right, you will "copy" the formula A1 + 10 as shown.

	A	B	C	D	E
1	100	100	100	100	100
2	A1+10				
3					
4					
5					
6					
7					
8					

	A	B	C	D	E
1	100	100	100	100	100
2	A1+10	B1+10	C1+10	D1+10	E1+10
3	A2+10				
4	A3+10				
5	A4+10				
6					
7					
8					

Row 2 is selected.   The spreadsheet fills the row and adjusts the formula.

Check It Out

Use the lower right spreadsheet above.

10. Select B1 to B5 and fill down. What number will be in B3?
11. Select A3 to C3 and fill right. What formula will be in C3? what number?
12. Select A4 to E4 and fill right. If D3 = 120, what formula will be in D4? what number?
13. Select E2 to E6 and fill down. If E5 = 140, what formula will be in E6? what number?

Spreadsheet Graphs

You can graph from a spreadsheet. As an example, let's use a spreadsheet to compare the **perimeter** of a square to the length of a side.

	A	B	C	D	E
1	side	perimeter			
2	1	4			
3	2	8			
4	3	12			
5	4	16			
6	5	20			
7	6	24			
8	7	28			
9	8	32			
10	9	36			
11	10	40			

File Edit

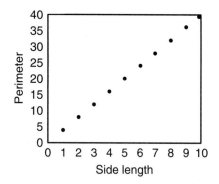

Most spreadsheets have a function that displays tables as graphs. See your spreadsheet reference for more information.

Check It Out

14. What cells gave the point $(1, 4)$?
15. What cells gave the point $(5, 20)$?
16. What point is shown by cells A9, B9?
17. What point is shown by cells A11, B11?

 **EXERCISES**

For the spreadsheet shown to the right, what number appears in each of the following cells?

1. B2   2. A3   3. C1

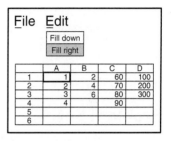

In which cell does each number appear?

4. 200   5. 3   6. 80

7. If the formula behind cell C2 is C1 + 10, what formula is behind cell C3?
8. What formula might be behind cell B2?
9. Say cells D5 and D6 were filled in using fill down from D3. D3 has the formula D2 + 100. What would the values of D5 and D6 be?
10. The formula behind cell B2 is A2 × 2. What formula might be behind cell B3?

Use the spreadsheet below to answer items 11–15.

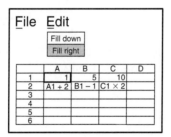

11. If you select A2 to A4 and fill down, what formula will appear in A4?
12. If you select C2 to C5 and fill down, what numbers will appear in C2 to C5?
13. If you select A2 to C2 and fill right, what will appear in B2? *Hint:* Fill will cover over any existing numbers or formulas.
14. If you select A1 to D1 and fill right, what will appear in D1?
15. If you select B2 to B5 and fill down (assuming the spreadsheet is as it appears above), what will appear in B4?

What have you learned?

You can use the problems and the list of words that follow to see what you have learned in this chapter. You can find out more about a particular problem or word by referring to the boldfaced topic number (for example, **9•2**).

Problem Set

Use your calculator to answer items 1–6. **9•1**
1. $65 + 12 \times 3 + 6^2$
2. 250% of 740

Round answers to the nearest tenth.
3. $17 - 5.6 \times 8^2 + 24$
4. $122 - (-45) \div 0.75 - 9.65$

5. Find the perimeter of rectangle *WXYZ*.
6. Find the area of rectangle *WXYZ* to the nearest tenth centimeter.

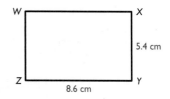

Use a scientific calculator to answer items 7–12. Round decimal answers to the nearest hundredth. **9•2**
7. 3.6^4
8. Find the reciprocal of 7.1.
9. Find the square of 7.5.
10. Find the square root of 7.5.
11. $(6 \times 10^5) \times (9 \times 10^4)$
12. $2.6 \times (13 \times 5.75)$

13. What is the measure of $\angle TRV$? **9•3**
14. What is the measure of $\angle VRS$? **9•3**
15. What is the measure of $\angle TRS$? **9•3**
16. Does $\overrightarrow{RT}$ divide $\angle SRV$ into two equal angles? **9•3**

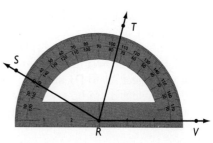

17. What would the measure be of an angle bisecting a right angle? **9•3**

For items 18–20, refer to the spreadsheet below. **9•4**

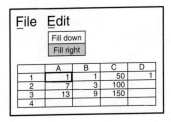

18. Name the cell holding 3.
19. A formula for cell C3 is C2 + 50. Name another formula for cell C3.
20. Cell D1 contains the number 1 and no formula. After using the command fill down, what number will be in cell D4?

WRITE DEFINITIONS FOR THE FOLLOWING WORDS.

hot **words**

angle **9•3**
arc **9•3**
cell **9•4**
circle **9•1**
column **9•4**
cube **9•2**
cube root **9•2**
decimal **9•1**

distance **9•3**
factorial **9•2**
formula **9•4**
horizontal **9•4**
negative number **9•1**
parentheses **9•2**
percent **9•1**
perimeter **9•4**
pi **9•1**
point **9•3**

power **9•2**
radius **9•1**
ray **9•3**
reciprocal **9•2**
root **9•2**
row **9•4**
spreadsheet **9•4**
square **9•2**
square root **9•1**
vertex **9•3**
vertical **9•4**

hot solutions

Solutions

Index

Chapter 1
Numbers and Computation

1. 30,000   **2.** 30,000,000

3. $(2 \times 10,000) + (4 \times 1,000) + (3 \times 100) + (7 \times 10) + (8 \times 1)$ **4.** 566,418; 496,418; 56,418; 5,618   **5.** 52,564,760; 52,565,000; 53,000,000

6. 0   **7.** 15   **8.** 3,589   **9.** 0

10. 400   **11.** 1,600

12. $(4 + 6) \times 5 = 50$ **13.** $(10 + 14) \div (3 + 3) = 4$

14. No   **15.** No   **16.** No   **17.** Yes

18. 3×11 **19.** $3 \times 5 \times 7$ **20.** $2 \times 2 \times 3 \times 3 \times 5$

21. 15   **22.** 7   **23.** 6

24. 15   **25.** 24   **26.** 80

27. 48

28. 6, 6   **29.** 13, −13   **30.** 15, 15   **31.** 25, −25

32. 6   **33.** −1   **34.** −18   **35.** 6   **36.** 0   **37.** 2

38. 28   **39.** −4   **40.** 7   **41.** 36   **42.** −30   **43.** −60

44. It will be a negative integer.

45. It will be a positive integer.

1·1 Place Value of Whole Numbers

1. 30   **2.** 3,000,000

3. Forty million, three hundred six thousand, two hundred   **4.** Fourteen trillion, thirty billion, five hundred million

5. $(8 \times 10,000) + (3 \times 1,000) + (4 \times 10) + (6 \times 1)$ **6.** $(3 \times 100,000) + (2 \times 100) + (8 \times 10) + (5 \times 1)$

p. 72　　**7.** $<$　**8.** $>$
　　　　　　9. 6,520; 52,617; 56,302; 526,000
　　　　　　10. 32,400　**11.** 560,000　**12.** 2,000,000
　　　　　　13. 400,000

1·2 Properties

p. 74　　**1.** Yes　**2.** No　**3.** No　**4.** Yes
p. 75　　**5.** 24,357　**6.** 99　**7.** 0　**8.** 1.5
　　　　　　9. $(3 \times 3) + (3 \times 6)$　**10.** $5 \times (8 + 7)$

1·3 Order of Operations

p. 78　　**1.** 10　**2.** 54

1·4 Factors and Multiples

p. 80　　**1.** 1, 2, 3, 6　**2.** 1, 2, 3, 6, 9, 18
p. 81　　**3.** 1, 2, 4　**4.** 1, 5
　　　　　　5. 2　**6.** 10
p. 82　　**7.** Yes　**8.** No　**9.** Yes　**10.** Yes
p. 84　　**11.** Yes　**12.** No　**13.** Yes　**14.** No
　　　　　　15. $2 \times 3 \times 5$　**16.** $2 \times 2 \times 2 \times 2 \times 5$ or $2^4 \times 5$
　　　　　　17. $2 \times 2 \times 2 \times 3 \times 5$ or $2^3 \times 3 \times 5$
　　　　　　18. $2 \times 5 \times 11$
p. 85　　**19.** 3　**20.** 10　**21.** 6　**22.** 12
p. 86　　**23.** 18　**24.** 50　**25.** 56　**26.** 150

1·5 Integer Operations

p. 88　　**1.** -3　**2.** $+250$
p. 89　　**3.** 12, 12　**4.** 4, -4　**5.** 8, 8　**6.** 0, 0
　　　　　　7. -2　**8.** 0　**9.** $+2$　**10.** -3
p. 90　　**11.** 6　**12.** -3　**13.** 5　**14.** -54
　　　　　　Oops!　If a can be positive, negative, or zero, $2 + a$
　　　　　　can be greater than, equal to, or less than 2.

Chapter 2
Fractions, Decimals, and Percents

p. 96   **1.** $45.75   **2.** $252   **3.** 92%   **4.** $5.55   **5.** C. $\frac{12}{21}$
     6. $4\frac{1}{6}$ **7.** $1\frac{9}{20}$ **8.** $3\frac{5}{6}$ **9.** $4\frac{37}{63}$
     10. B. $2\frac{1}{2}$ **11.** $\frac{5}{16}$ **12.** $\frac{9}{65}$ **13.** $\frac{1}{2}$ **14.** $\frac{144}{217}$
     15. Hundredths   **16.** $4 + 0.6 + 0.003$ **17.** 0.247

p. 97   **18.** 1.065; 1.605; 1.655; 16.5
     19. 17.916   **20.** 13.223   **21.** 101.781
     22. 25%   **23.** 5.3   **24.** 19.7%
     25. 68%   **26.** 50%
     27. 6%   **28.** 56%
     29. 0.34   **30.** 1.25
     31. $\frac{7}{25}$ **32.** $1\frac{3}{10}$

2·1 Fractions and Equivalent Fractions

p. 99   **1.** $\frac{2}{3}$ **2.** $\frac{4}{9}$ **3.** Answers will vary.

p. 100   **4–7.** Answers will vary.

p. 102   **8.** =   **9.** =   **10.** ≠

p. 103   **11.** $\frac{1}{5}$ **12.** $\frac{1}{3}$ **13.** $\frac{9}{10}$

p. 104   **Musical Fractions** $\frac{1}{32}, \frac{1}{64}$;

$$\mathbf{o} = \text{♩♩} = \text{♩♩♩♩} = \text{♫♫♫♫} = \text{♫♫♫♫}$$

p. 106   **14.** $4\frac{4}{5}$ **15.** $1\frac{4}{9}$ **16.** $2\frac{3}{4}$ **17.** $4\frac{5}{6}$
     18. $\frac{17}{10}$ **19.** $\frac{41}{8}$ **20.** $\frac{33}{5}$ **21.** $\frac{52}{7}$

2·2 Comparing and Ordering Fractions

p. 109   **1.** <   **2.** <   **3.** >   **4.** =
     5. >   **6.** >   **7.** >

p. 110   **8.** $\frac{2}{4}; \frac{5}{8}; \frac{4}{5}$ **9.** $\frac{7}{12}; \frac{2}{3}; \frac{3}{4}$ **10.** $\frac{5}{8}; \frac{2}{3}; \frac{5}{6}$

2.3 Addition and Subtraction of Fractions

p. 112 **1.** 2 **2.** $\frac{6}{25}$ **3.** $\frac{5}{23}$ **4.** $\frac{3}{8}$

p. 114 **5.** $1\frac{1}{4}$ **6.** $\frac{1}{6}$ **7.** $\frac{7}{10}$ **8.** $\frac{7}{12}$

9. $9\frac{5}{6}$ **10.** $34\frac{5}{8}$ **11.** 61

p. 115 **12.** $6\frac{3}{8}$ **13.** $60\frac{1}{6}$ **14.** $59\frac{3}{8}$

p. 116 **15.** $6\frac{3}{8}$ **16.** $16\frac{1}{6}$ **17.** $17\frac{29}{30}$ **18.** $8\frac{7}{10}$

p. 117 **19.** $3\frac{1}{2}$ or 3 **20.** 7 **21.** 9 **22.** $7\frac{1}{2}$ or 8

p. 118 **The Ups and Downs of Stocks** 1%

2.4 Multiplication and Division of Fractions

p. 121 **1.** $\frac{5}{18}$ **2.** $\frac{15}{28}$ **3.** $\frac{2}{9}$ **4.** $\frac{33}{100}$

5. $\frac{1}{6}$ **6.** $\frac{3}{7}$ **7.** $\frac{16}{45}$ **8.** $\frac{2}{3}$

p. 122 **9.** $\frac{8}{3}$ **10.** $\frac{1}{5}$ **11.** $\frac{2}{9}$

p. 123 **12.** 8 **13.** $9\frac{19}{48}$ **14.** $72\frac{11}{24}$ **15.** $78\frac{3}{8}$

16. $1\frac{1}{4}$ **17.** $1\frac{3}{7}$ **18.** $6\frac{2}{9}$

p. 124 **19.** $1\frac{1}{2}$ **20.** $5\frac{1}{3}$ **21.** $\frac{3}{4}$

2.5 Naming and Ordering Decimals

p. 127 **1.** 0.9 **2.** 0.55 **3.** 7.18 **4.** 5.03

5. 0.6 + 0.03 + 0.004 **6.** 3 + 0.2 + 0.02 + 0.001

7. 0.07 + 0.007

p. 129 **8.** Five ones; five and six hundred thirty-three thousandths **9.** Five thousandths; forty-five thousandths **10.** Seven thousandths; six and seventy-four ten thousandths **11.** One hundred thousandth; two hundred seventy-one hundred thousandths

12. < **13.** < **14.** >

p. 130 **15.** 4.0146; 4.1406; 40.146 **16.** 8; 8.073; 8.373; 83.037 **17.** 0.52112; 0.522; 0.5512; 0.552

18. 1.66 **19.** 226.95 **20.** 7.40 **21.** 8.59

HOT SOLUTIONS

2•6 Decimal Operations

p. 132   **1.** 88.88   **2.** 61.916   **3.** 6.13   **4.** 46.283
p. 133   **5.** 12   **6.** 4   **7.** 14   **8.** 13
p. 134   **9.** 4.704   **10.** 114.1244
p. 135   **11.** 0.001683   **12.** 0.048455
p. 136   **13.** 210   **14.** 400 Answers may vary.
Olympic Decimals   9.52; 9.7
p. 138   **15.** 5   **16.** 4.68   **17.** 50.4   **18.** 46
19. 0.73   **20.** 0.26   **21.** 0.60

2•7 Meaning of Percent

p. 140   **1.** 32% shaded; 68% not shaded   **2.** 44% shaded; 56% not shaded   **3.** 15% shaded; 85% not shaded
p. 141   **4.** 23   **5.** 40   **6.** 60   **7.** 27
p. 142   **8.** $1.45   **9.** $4   **10.** $50 Answers may vary.
Luxuries or Necessities?   About 923,000,000

2•8 Using and Finding Percents

p. 144   **1.** 19.25   **2.** 564   **3.** 12.1   **4.** 25.56
p. 145   **5.** 36.4   **6.** 11.16   **7.** 93.13   **8.** 196
p. 146   **9.** 25%   **10.** 15%   **11.** 5%   **12.** 150%
p. 147   **13.** 108   **14.** 20   **15.** 23.33   **16.** 925
p. 148   **17.** 93%   **18.** 178%   **19.** 1,367%   **20.** 400%
p. 149   **21.** 68%   **22.** 57%   **23.** 22%   **24.** 58%
p. 150   **25.** Discount: $18.75; Sale price: $56.25
26. Discount: $27; Sale price: $153
p. 151   **27.** 100   **28.** 2   **29.** 15   **30.** 30 Answers may vary
p. 152   **31.** $I = $214.50, A = 864.50
32. $I = $840.00, A = $3,240.00$

2·9 Fraction, Decimal, and Percent Relationships

p. 155   **1.** 55% **2.** 40% **3.** 75% **4.** 43%
5. $\frac{4}{25}$ **6.** $\frac{1}{25}$ **7.** $\frac{19}{50}$ **8.** $\frac{18}{25}$

p. 156   **9.** $\frac{49}{200}$ **10.** $\frac{67}{400}$ **11.** $\frac{969}{800}$ or $1\frac{169}{800}$

Honesty Pays   20%

p. 157   **12.** 45% **13.** 60.6% **14.** 1.9% **15.** 250%

p. 158   **16.** 0.54 **17.** 1.9 **18.** 0.04 **19.** 0.29

p. 159   **20.** 0.8 **21.** 0.3125 **22.** $0.5\overline{5}$ or $0.\overline{5}$

p. 160   **23.** $\frac{9}{40}$ **24.** $\frac{43}{80}$ **25.** $\frac{9}{25}$

Chapter 3   Powers and Roots

p. 166   **1.** 7^5 **2.** a^8 **3.** 4^3 **4.** x^2 **5.** 3^4
6. 4 **7.** 25 **8.** 100 **9.** 49 **10.** 144
11. 8 **12.** 64 **13.** 1,000 **14.** 343 **15.** 1
16. 100 **17.** 1,000,000 **18.** 10,000,000,000
19. 10,000,000 **20.** 10

p. 167   **21.** 3 **22.** 5 **23.** 12 **24.** 8 **25.** 2
26. 4 and 5 **27.** 6 and 7 **28.** 2 and 3 **29.** 8 and 9
30. 1 and 2
31. 2.236 **32.** 4.472 **33.** 7.071 **34.** 9.110
35. 7.280

3·1 Powers and Exponents

p. 168   **1.** 8^4 **2.** 3^7 **3.** x^3 **4.** y^5

p. 169   **5.** 16 **6.** 25 **7.** 64 **8.** 36

p. 170   **9.** 27 **10.** 216 **11.** 729 **12.** 125

p. 171   **13.** 1,000 **14.** 100,000 **15.** 10,000,000,000
16. 100,000,000

p. 172   **Bugs** 1.2×10^{18}—that is, one quintillion, two
hundred quadrillion

3•2 Square Roots

p. 174 **1.** 4 **2.** 5 **3.** 8 **4.** 10

p. 175 **5.** Between 4 and 5 **6.** Between 6 and 7

p. 176 **7.** 1.414 **8.** 5.292

Chapter 4
Data, Statistics, and Probability

p. 182 **1.** No **2.** Unbiased

3. Bar graph **4.** Wednesday **5.** You cannot tell from this graph.

6.

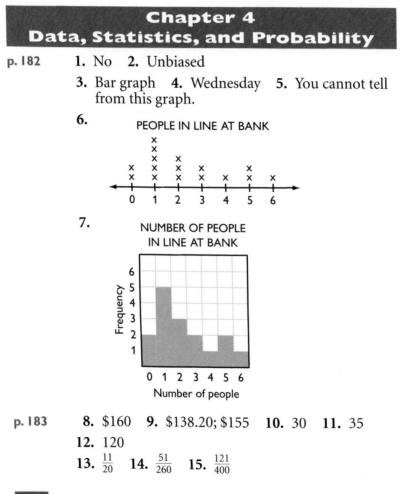

PEOPLE IN LINE AT BANK

7.

NUMBER OF PEOPLE
IN LINE AT BANK

p. 183 **8.** $160 **9.** $138.20; $155 **10.** 30 **11.** 35

12. 120

13. $\frac{11}{20}$ **14.** $\frac{51}{260}$ **15.** $\frac{121}{400}$

4•1 Collecting Data

p. 184 **1.** Students signed up for after-school sports; 60

2. Wolves on Isle Royale; 15

p. 185   **3.** Answers will vary.   **4.** No; it is limited to people who are in that fitness center, so they may like it best.

p. 186   **5.** It calls table tennis tame.   **6.** It does not imply that you are more adventurous if you like the sport.   **7.** Do you donate money to charity?

p. 187   **8.** 3   **9.** Yes

10. Have a party; most students said yes.

4·2 Displaying Data

p. 190   **1.** 18

2.

Number of Sponsors	2	3	4	5	6	7	8	9	10
Number of Students	3	0	3	5	2	0	2	3	4

p. 192   **3.** 52   **4.** 25%   **5.** 50%

p. 193   **6.** Nonfiction and videos   **7.** Possible answer: The library has slightly more nonfiction than fiction in its collection.

8. PET SHOW ENTRIES

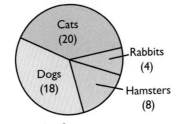

p. 194   **9.** 5   **10.** 2

11.

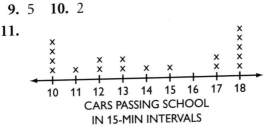

CARS PASSING SCHOOL
IN 15-MIN INTERVALS

p. 195   **12.** 1975   **13.** False

14. Possible answer: 1970 and 1980

p. 196   **15.** 20   **16.** Twenties   **17.** 28

p. 198   **18.** Bristol   **19.** Answers will vary.

20.

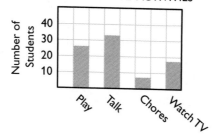

AFTER-SCHOOL ACTIVITIES

21. 60   **22.** Answers will vary.

p. 199   **Graphic Impressions**   One might think the size of the pictures represents the *size* of the animals; the bar graph more accurately portrays the data.

p. 200   **23.** 16

24.

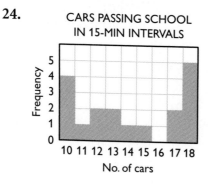

CARS PASSING SCHOOL
IN 15-MIN INTERVALS

4•3 Analyzing Data

p. 202   **1.** Yes   **2.** Yes   **3.** No

4•4 Statistics

p. 205   **1.** 24.75   **2.** 87   **3.** 228   **4.** $37
p. 206   **5.** 18   **6.** 25   **7.** 96   **8.** 157.5 lb
p. 207   **9.** 53   **10.** 96 and 98   **11.** 14   **12.** No mode
p. 208   **13.** 613   **14.** 75   **15.** 32°   **16.** 384

4·5 Combinations and Permutations

1. 9   **2.** 12   **3.** 8   **4.** 8

5. 60   **6.** 30   **7.** 720   **8.** 132   **9.** 5,040   **10.** 120

11. 21   **12.** 210   **13.** 56

14. 24 times as many permutations as combinations

Monograms   17,576 monograms

4·6 Probability

1. $\frac{3}{40}$ **2.** $\frac{11}{20}$ **3.** Answers will vary.

4. $\frac{1}{2}$ **5.** $\frac{1}{3}$ **6.** 1   **7.** $\frac{5}{12}$

8. $\frac{1}{4}$; 0.25; 1:4; 25%   **9.** $\frac{2}{5}$; 0.4; 2:5; 40%

10. 2T, 4H   **11.** Answers will vary.

Lottery Fever   Struck by lightning; $\frac{260}{260,000,000}$ is about 1 in 1 million, compared to the 1-in-16-million chance of winning a 6-out-of-50 lottery.

12.

	1	2	3
1	11	12	13
2	21	22	23
3	31	32	33

13. $\frac{1}{3}$

14.   **15.**

16. $\frac{1}{12}$; independent   **17.** $\frac{18}{95}$; dependent

18. $\frac{1}{4}$ **19.** $\frac{9}{38}$

How Mighty Is the Mississippi?   2,949.5 mi; 2,620 mi; 1,830 mi

Chapter 5 Logic

1. False   **2.** True   **3.** True   **4.** False   **5.** True

6. If an angle is a right angle, then it has a measure of 90°.   **7.** If a triangle is acute, then it has three acute angles.

Continued

HOT SOLUTIONS

p. 234 (cont.)

8. If $2 \times n = 16$, then $n = 8$. **9.** If I wear my white shoes, then it is summertime.

10. Adam did not get the highest score on the math test. **11.** This triangle is equilateral.

12. If you do not study, then you will not receive a good grade. **13.** If $a + 1 \neq 6$, then $a \neq 5$.

p. 235

14. If you do not pay the full admission price, then you are not over 12 years old. **15.** If the area formula of a figure is not $A = (\frac{1}{2})bh$, then the figure is not a triangle.

16. February **17.** 2 and 14

18. $\{2\}, \{4\}, \{2, 4\}, \emptyset$ **19.** $\{2\}, \{4\}, \{6\}, \{2, 4\}, \{2, 6\}, \{4, 6\}, \{2, 4, 6\}, \emptyset$

20. $\{5, 7, 9, 11\}$ **21.** $\{a, b, c, d, e, g\}$ **22.** $\{2, 9\}$

23. $\{x, y, z\}$ **24.** $\{3, 4, 5, 6, 7, 8\}$

25. $\{8, 16\}$ **26.** $\emptyset$

27. $\{1, 2, 3, 4, 5\}$ **28.** $\{5, 6, 7, 8, 9\}$

29. $\{1, 2, 3, 4, 5, 6, 7, 8, 9\}$ **30.** $\{5\}$

5·1 If/Then Statements

p. 237

1. If a whole number ends with a 2, then the number is even. **2.** If a polygon has 3 sides, then it is a triangle.

3. If you got a product of 12, then you multiplied 3 and 4. **4.** If an angle is a right angle, then it has a measure of 90°.

p. 238

5. We will not go on the class trip. **6.** 3 is not less than 4.

7. If an integer does not end with 0, then you cannot divide it by 10. **8.** If today is not Tuesday, then tomorrow is not Wednesday.

p. 239

9. If you did not get a product of 42, then you did not multiply 6 and 7. **10.** If two lines do cross, then they are not parallel.

p. 240   **Who's Who?** Tanya is in the third seat, Sylvia is in the middle seat, and Leslie is lying. Since Tanya always tells the truth, she can't be the girl in the middle (otherwise she would say her name is Tanya). That also means the girl in the first seat is lying. So Tanya must be in the third seat, telling the truth that Sylvia is in the middle. The girl in the first seat must be Leslie.

5·2 Counterexamples

p. 242   **1.** True; False; counterexample: 6   **2.**   True; True

5·3 Sets

p. 244   **1.** False   **2.** True

       3. $\{4\}, \{8\}, \{4, 8\}, \varnothing$ **4.** $\{m\}, \varnothing$

p. 245   **5.** $\{1, 2, 7, 8\}$ **6.** $\{5, 10\}$ **7.** $\{12\}$ **8.** $\varnothing$

p. 246   **9.** $\{1, 4, 5\}$ **10.** $\{5\}$ **11.** $\{1, 4, 5, 6\}$ **12.** $\{1, 5\}$

       13. $\{5\}$

Chapter 6   Algebra

p. 252   **1.** $x + 5$ **2.** $4n$ **3.** $2(x - 6)$ **4.** $\frac{n}{3} - 2$

       5. $2x - 3 = x + 5$ **6.** $6(n + 2) = 2n - 4$

       7. $6(x + 3)$ **8.** $5(2n - 3)$ **9.** $3(a - 7)$

       10. $3x + 7$ **11.** $8a + 2b$ **12.** $7n - 10$

       13. 15 mi   **14.** 10 boys   **15.** 7.5 cm

p. 253   **16.**

$x < 3$

       17.

$x \geq 4$

       18.

$n > 2$

Continued

HOT SOLUTIONS

p. 253
(cont.)

19.

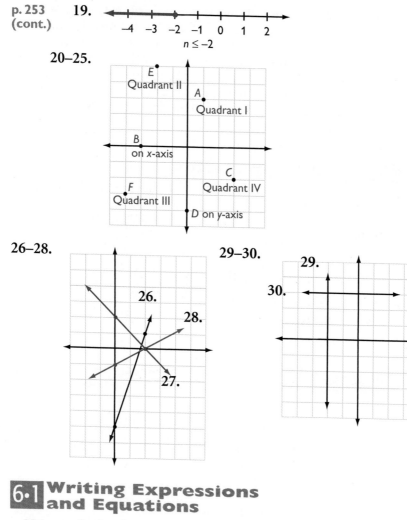

$n \leq -2$

20–25.

26–28.

29–30.

6·1 Writing Expressions and Equations

p. 254   **1.** 2 **2.** 1 **3.** 3 **4.** 2

p. 255   **5.** $5 + x$ **6.** $n + 4$ **7.** $y + 8$ **8.** $n + 2$

p. 256   **9.** $8 - x$ **10.** $n - 3$ **11.** $y - 6$ **12.** $n - 4$

p. 257   **13.** $4x$ **14.** $8n$ **15.** $0.25y$ **16.** $9n$

p. 258   **17.** $\frac{x}{5}$ **18.** $\frac{8}{n}$ **19.** $\frac{20}{y}$ **20.** $\frac{a}{4}$

p. 259   **21.** $4n - 6$ **22.** $\frac{8}{x} - 5$ **23.** $2(n - 4)$

Three Astronauts and a Cat   79 fish

p. 260   **24.** $5x - 9 = 6$ **25.** $\frac{x}{3} + 6 = x - 2$
          26. $4n - 1 = 2(n + 5)$

6•2 Simplifying Expressions

p. 262   **1.** No   **2.** Yes   **3.** No   **4.** Yes
          5. $7 + 2x$ **6.** $6n$ **7.** $4y + 5$ **8.** $8 \cdot 3$

p. 263   **...3, 2, 1 Blast Off**   Answers will vary depending on height. To match a flea's feat, a 5-foot-tall child would have to jump 800 feet.

p. 264   **9.** $4 + (5 + 8)$ **10.** $2 \cdot (3 \cdot 5)$ **11.** $5x + (4y + 3)$
          12. $(6 \cdot 9)n$
          13. $6(100 - 1) = 594$ **14.** $3(100 + 6) = 318$
          15. $4(200 - 2) = 792$ **16.** $5(200 + 10 + 1) = 1{,}055$

p. 265   **17.** $6x + 2$ **18.** $12n - 18$ **19.** $-8y + 2$
          20. $10x - 8$

p. 266   **21.** $7(x + 2)$ **22.** $2(2n - 5)$ **23.** $10(c + 5)$
          24. $9(2a - 3)$

p. 268   **25.** $11x$ **26.** $6y$ **27.** $8n$ **28.** $-6a$
          29. $2y + 4z$ **30.** $5x - 6$ **31.** $5a + 2$ **32.** $7n + 2$

6•3 Evaluating Expressions and Formulas

p. 270   **1.** 11   **2.** 9   **3.** 9   **4.** 14
p. 271   **5.** 32 cm   **6.** 18 ft
p. 272   **7.** 30 mi   **8.** 1,200 km   **9.** 220 mi   **10.** 6 ft

6•4 Ratio and Proportion

p. 274   **1.** $\frac{4}{8} = \frac{1}{2}$ **2.** $\frac{8}{12} = \frac{2}{3}$ **3.** $\frac{12}{4} = \frac{3}{1} = 3$
p. 275   **4.** Yes   **5.** No
p. 276   **6.** 3.5 gal   **7.** \$105

6•5 Inequalities

p. 279 **1.**
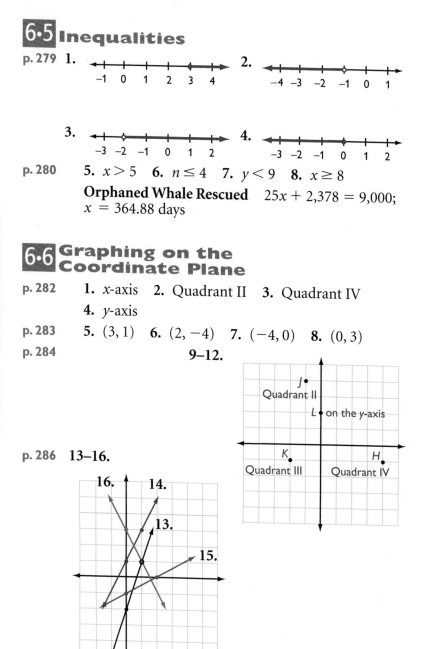
2.

3. **4.**

p. 280 **5.** $x > 5$ **6.** $n \leq 4$ **7.** $y < 9$ **8.** $x \geq 8$

Orphaned Whale Rescued $25x + 2{,}378 = 9{,}000;$
$x = 364.88$ days

6•6 Graphing on the Coordinate Plane

p. 282 **1.** x-axis **2.** Quadrant II **3.** Quadrant IV
4. y-axis

p. 283 **5.** $(3, 1)$ **6.** $(2, -4)$ **7.** $(-4, 0)$ **8.** $(0, 3)$

p. 284 **9–12.**

Quadrant II
J
L on the y-axis
K
H
Quadrant III   Quadrant IV

p. 286 **13–16.**

16. **14.**
13.
15.

p. 287   **17–20.**

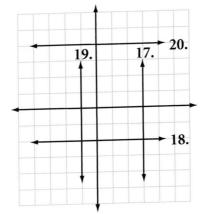

19.   17.   **20.**

18.

Chapter 7 Geometry

p. 294

1. Possible answers: $\angle ABC$ or $\angle CBA$; $\angle ABD$ or $\angle DBA$; $\angle DBC$ or $\angle CBD$ **2.** Possible answers: $\overrightarrow{BA}$, $\overrightarrow{BD}$, $\overrightarrow{BC}$ **3.** Right angle; 90° angle

4. 45°

5. Parallelogram   **6.** 125°   **7.** 360°

8. 16 ft   **9.** 30 mm   **10.** 84 cm^2

11. 54 in.^{2} **12.** 216 mm^2 **13.** 256 in.^{2}

p. 295

14. B   **15.** A   **16.** C

17. 60 cm^3 **18.** 78.4 mm^3

19. 80°   **20.** 28.3 in.^{2}

7·1 Naming and Classifying Angles and Triangles

p. 296   **1.** $\overleftrightarrow{MN}$, $\overleftrightarrow{NM}$ **2.** $\overrightarrow{QR}$, $\overrightarrow{QS}$

p. 298   **3.** N **4.** $\angle MNO$ or $\angle ONM$; $\angle MNP$ or $\angle PNM$; $\angle ONP$ or $\angle PNO$

p. 299   **5.** 40° **6.** 105° **7.** 140°

p. 300   **8.** 25°, acute **9.** 90°, right **10.** 115°, obtuse

p. 302   **11.** $m\angle J = 80°$ **12.** $m\angle E = 70°$ **13.** C **14.** A

7·2 Naming and Classifying Polygons and Polyhedrons

p. 305   **1.** Possible answers: *OLMN, ONML, LMNO, LONM, MNOL, MLON, NOLM, NMLO*

2. 265° **3.** 95°

p. 306   **4.** Square **5.** Each angle measures 90°.
6. Each side measures 3 in.

p. 308   **7.** Hexagon **8.** Pentagon **9.** Rectangle

10. Octagon

Oh, Obelisk! 50°

p. 310   **11.** 900° **12.** 108°

p. 311   **13.** Square pyramid; Rectangular pyramid

7·3 Symmetry and Transformations

p. 315   **1.**

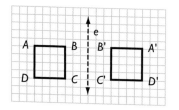

Fish Farming 15,000 m²

p. 316   **2.** B, C, D, E, K, M, T, U, V, W, Y **3.** H, I, O

p. 317   **4.** 90° **5.** 180°

p. 318   **6.** *B, D, F*

7·4 Perimeter

p. 321   **1.** 360 in. **2.** 65 ft **3.** 116 in. **4.** 24 km

p. 322   **5.** 24 cm **6.** 30 ft

7·5 Area

1. C **2.** B **3.** About 50 in.^{2}

4. 408 yd^2 **5.** 81 in.^{2}

6. 180 cm^2 **7.** 8 ft

8. 64 ft^2 **9.** 24 cm^2

10. 26 m^2 **11.** 68 cm^2

7·6 Surface Area

1. 72 in.^{2} **2.** 384 cm^2

3. 84 ft^2 **4.** 251.2 ft^2

7·7 Volume

1. 4 in.^{3} **2.** 4 in.^{3}

3. 216 in.^{3} **4.** 90 m^3

5. 785 m^3 **6.** 2,813.4 yd^3

7. 440 m^3 **8.** 94.2 in.^{3}

Good Night, T. Rex About 1,175,000 mi^3

7·8 Circles

1. 6 cm **2.** 7.5 ft **3.** $\frac{y}{2}$

4. 20 in. **5.** 11 m **6.** $2x$

7. 8π mm **8.** 31.4 m **9.** 14 ft **10.** 9 cm

11. ∠XYZ **12.** 45° **13.** 90° **14.** 360°

Around the World About $2\frac{2}{5}$ times

15. 49π mm^2; 153.9 mm^2 **16.** 201 ft^2

Chapter 8 Measurement

1. Centiliter **2.** Kilogram **3.** Millimeter

4. 10 **5.** 8,000 **6.** 6 **7.** 45

8. 78 **9.** 936 **10.** 26

11. 360 **12.** 51,840 **13.** 40

p. 350
(cont.)
14. 20,000 **15.** 72 **16.** 27 **17.** 1,000 **18.** 1
19. 8

p. 351 **20.** 17,280 in.^{3} **21.** 10 ft^3 **22.** $0.43 per pound; the per pound rate would cost $8.72 less than the cubic foot rate. **23.** Yes
24. 2 **25.** $\frac{9}{4}$

8·1 Systems of Measurement

p. 353 **1.** Metric **2.** Customary
p. 354 **3.** 6.25, 6.34 **4.** 39, 40

8·2 Length and Distance

p. 356 **1–2.** Answers will vary.
p. 357 **3.** 6 ft **4.** 360 in. **5.** 1.5 m **6.** 5 cm
p. 358 **7.** 3.9 **8.** 3.2 **9.** A **10.** C **11.** A

8·3 Area, Volume, and Capacity

p. 360 **1.** 500 **2.** 432
p. 361 **3.** 2.9 ft^3 **4.** 64,000 mm^3
p. 362 **5.** 8 **6.** $\frac{1}{2}$

8·4 Mass and Weight

p. 364 **1.** 2 **2.** 7,000 **3.** 80 **4.** 2,500 **5.** 52
p. 365 **Poor Sid** No, the mass is always the same.

8·5 Time

p. 366 **1.** 252 months **2.** September 10, 2013
p. 367 **The World's Largest Reptile** 5 times as long

8·6 Size and Scale

p. 368 **1.** *A* and *D*
p. 369 **2.** 3 **3.** $\frac{1}{4}$
p. 370 **4.** $\frac{1}{16}$ **5.** 8 in. × 12 in.

Chapter 9 Tools

p. 376　**1.** 91　**2.** 1,350　**3.** 116.8　**4.** −10.3　**5.** 25 cm
6. 34 cm² **7.** 166.38　**8.** 0.13　**9.** 153.76　**10.** 2.12
11. 1.8×10^6　**12.** 168.75

p. 377　**13.** 45°　**14.** 135°　**15.** 90°　**16.** No
17. Compass, straightedge
18. D1　**19.** A2 × 2　**20.** 1,000

9·1 Four-Function Calculator

p. 379　**1.** 15.8　**2.** 31.5　**3.** −30.　**4.** −48.

p. 380　**5.** 60　**6.** 69　**7.** 454　**8.** 400

p. 381　**Magic Numbers** $\frac{100x + 10x + x}{x + x + x} = \frac{111x}{3x} = 37$

p. 382　**9.** 20　**10.** 544　**11.** 14　**12.** 112.5

9·2 Scientific Calculator

p. 386　**1.** 4.5　**2.** 343　**3.** 5,040　**4.** 6,561　**5.** 3
6. 10.8　**7.** 0.5　**8.** 361　**9.** 2,068　**10.** −344

p. 387　**Calculator Alphabet** hello

9·3 Geometry Tools

p. 388　**1.** 1 in. or 2.5 cm　**2.** 5 cm or 2 in.　**3.** 3 in. or
7.6 cm　**4.** 2 cm or $\frac{3}{4}$ in.

p. 390　**5.** 130°　**6.** 45°

p. 391　**7–10.** To check answers,
measure each radius.

p. 393　**12.** Answers will vary.

p. 393　**11.**

9·4 Spreadsheets

p. 396　**1.** 5　**2.** 6　**3.** 4　**4.** True　**5.** False

p. 397　**6.** B3 × C3　**7.** B4 × C4　**8.** $10　**9.** $60

p. 399　**10.** 100　**11.** C2 + 10; 120　**12.** D3 + 10; 130
13. E5 + 10; 150

p. 400　**14.** A2, B2　**15.** A6, B6　**16.** (8, 32)　**17.** (10, 40)

HOT SOLUTIONS

INDEX

Index